Chemical Infrared
Fourier Transform Spectroscopy

CHEMICAL ANALYSIS

A SERIES OF MONOGRAPHS ON
ANALYTICAL CHEMISTRY AND ITS APPLICATIONS

VOLUME 43

A WILEY-INTERSCIENCE PUBLICATION

JOHN WILEY & SONS
New York/London/Sydney/Toronto

Chemical Infrared Fourier Transform Spectroscopy

PETER R. GRIFFITHS

Associate Professor of Chemistry
Ohio University
Athens, Ohio.

A WILEY-INTERSCIENCE PUBLICATION

JOHN WILEY & SONS

New York • Chichester • Brisbane • Toronto • Singapore

Library of Congress Cataloging in Publication Data

Griffiths, Peter R   1942–
   Chemical infrared Fourier transform spectroscopy.

   (Chemical analysis; v. 43)
   "A Wiley-Interscience publication."
   Includes bibliographical references and index.
   1. Infra-red spectrometry. 2. Fourier transform spectroscopy: I. Title. II. Series.

QD96.I5G74   544'.63   75–6505
ISBN 0–471–32786–7

Printed in the United States of America

10 9 8 7 6 5

PREFACE

In the last decade, the mathematical operation known as the Fourier transform has been applied to many different branches of chemistry. However, its
application to infrared and NMR spectroscopy has allowed the most spectacular advances to be made in these two areas. This book describes the theory,
instrumentation and chemical applications of *infrared* Fourier transform
spectroscopy (FTS). It has been written in a style that should make the
material readily understandable to a B.S. chemist. For example, several optical equations are derived in the first chapter using simple two-dimensional
optical ray diagrams which give results which are only slightly different from
the equations obtained using rigorous three-dimensional models.

The book is divided into two principal sections, the first on theory and
instrumentation and the second covering the chemical applications of FTS.
The Michelson interferometer is first described, and the generation of interferograms is discussed with respect to spectral resolution, instrument line
shape, apodization, phase correction and the effect of optical aberrations.
After chapters discussing the digital sampling of interferograms and the
digital Fourier transform, the way in which all these factors affect instrument
design is illustrated with respect to past and present Fourier transform
spectrometers; different chapters cover interferometers, and their auxiliary
optics and data systems. A chapter on the technique known as dual-beam
FTS describes one of the more important innovations in interferometry,
and the first section is concluded with a discussion of signal-to-noise ratio
in FTS.

Throughout the book, I have tried to illustrate many of the theoretical
concepts with practical examples. In the second section, published work performed with both far and mid infrared Fourier transform spectrometers is
described, and this section should give spectroscopists and non-spectros-
copists alike a better idea of the types of measurements that can be made
using this important tool.

Before finishing this preface, I would like to thank my many colleagues,
past and present, without whose help and guidance I would not have been in
the position to write this book. In particular, I would like to thank Professor
Sir Harold Thompson of St. John's College, Oxford University and the late
Professor Ellis Lippincott of Maryland University, who guided my early
career in infrared spectroscopy, and the scientific staff at Block Engineering
and Digilab with whom I had the privilege of working while many of the

innovations described in this book were being made. Finally, and most important, I would like to thank my wife, Dolly, for putting up with me while I was preparing the manuscript for this book.

PETER GRIFFITHS

Athens, Ohio
December 1974

CONTENTS

Chemical Infrared
Fourier Transform Spectroscopy

INTRODUCTION

Infrared spectrophotometry is probably the most widely used tool in the world today for the identification of organic compounds. Among the reasons for the tremendous popularity of the technique are its simplicity, versatility, accuracy, and cheapness. Spectra of gaseous, liquid, and solid samples can all be measured routinely with no special sample-handling technique needed on the part of the operator. Functional group information can be gathered easily from an infrared spectrum, and once the spectroscopist has a good idea as to the type of compound that he is measuring, a vast amount of reference data is available in the literature for direct comparison. Thus, unknown materials can usually be rapidly identified. Similarly, individual components of a mixture of known compounds can usually be quantified by infrared spectroscopy provided that the spectrum of each pure component is available.

In spite of all its apparent benefits, infrared spectroscopy, as most chemists know it, has certain drawbacks. Measurements are slow, and only specially designed spectrometers can measure the complete mid-infrared spectrum in less than a few minutes with acceptable resolution and signal-to-noise ratio. Linked to this time limitation is the fact that most infrared spectrometers are not able to measure spectra when the sample has a low baseline transmittance (less than 1%) without taking an excessively long measurement time.

More and more chemical problems are being encountered for which the sensitivity of the conventional grating spectrometer is not quite adequate. These applications may involve the measurement of very weak bands, the attainment of extremely high resolution, the need for very short measurement times, or monitoring very weak sources of infrared radiation. The development of Fourier transform spectroscopy has allowed many of these difficult measurements to be made much more easily. To understand where the benefits of Fourier transform spectroscopy arise, it is first necessary to understand where the limitations of conventional spectrometers originate.

The fundamental drawback of prism or grating spectrometers is found in the central component—the monochromator. Narrow slits are situated at both the entrance and the exit of the monochromator which limit the frequency range of the radiation reaching the detector to one resolution width. The purpose of the entrance slit is to define the beam of light entering the monochromator so that a sufficiently collimated beam of light is incident on the prism or grating. After dispersion of the beam, the second (exit) slit allows only the one small frequency element of interest to reach the detector. The remainder of the beam is reflected from, or absorbed by, the jaws of the slit and does not reach the detector. The narrower the width of the entrance and

1

exit slits, the higher is the frequency resolution of the spectrometer. However, as the resolution is increased, an even greater proportion of the total radiation from the source is "wasted" on the jaws of the exit slit, and less energy reaches the detector. Even for a low resolution grating spectrometer operating between 4000 and 400 cm^{-1} at a resolution of 8 cm^{-1}, the proportion of energy reaching the detector at any instant is 8:3600, or about 0.2%. When the resolution is increased to 1 cm^{-1}, the amount of radiation not reaching the detector is 99.97%.

It is therefore easily seen that the monochromator is an inefficient device for measuring infrared spectra over wide frequency ranges at any reasonable resolution. However, until recently no way of coding the spectral information other than the dispersion by prisms or gratings was considered feasible for the economical measurement of infrared spectra.

At the end of the last century, a different technique had been studied by Michelson, who was the inventor of the *interferometer* which now bears his name. The phenomenon of interference of light had been recognized long before this time, but among the many significant contributions to the field of optics made by Michelson [1, 2] was the design of an instrument in which two interfering beams are well separated in space so that their relative path differences may be conveniently and precisely varied. Michelson has also been credited as the first person to name such a device an "interferometer." Although he was not able to measure a well-defined interference pattern (interferogram) as we know them today, Michelson apparently realized that spectral information could be derived from the envelope, or visibility curve, of the interference fringes generated by his interferometer [3]. From one of these visibility curves, Michelson was able to postulate that the red Balmer line in the spectrum of hydrogen was not a singlet but a doublet which at that time nobody had been able to resolve. Subsequent work confirmed his suggestion.

In 1892, Lord Rayleigh [4] recognized that the interferogram was related to the spectrum of the radiation passing through an interferometer through the mathematical operation known as the Fourier transformation. Both Michelson and Lord Rayleigh realized that the measurement of the complete interferogram was beyond the technical resources of the day and that their work would have to be performed with the visibility curve alone.

In an attempt to calculate spectral data from his measured interference patterns, Michelson conceived and built an ingenious special purpose "computer." This was an 80-channel analog device consisting of 80 gears driving 80 wheels to rotate at speeds proportional to the integers 1 through 80. Each wheel rocked a lever which in turn generated a simple harmonic motion in an adjustable arm. Each arm was linked by springs to an axle, and the composite motion of all 80 arms moved a pen. There is actually no record of Michelson ever having used this device successfully with a measured interference pattern, but he did successfully reinvert a synthetic visibility curve [3].

The first measurement of a well-resolved interferogram with a two-beam interferometer was performed in 1910 by Rubens, Hollnagel and Wood [5, 6], but again the technology required for the full Fourier transformation was beyond these workers at that time. Like Michelson, they guessed at plausible spectral distributions, calculated the interferograms from these spectra, and compared them with their measured interferograms.

The first person to actually perform a numerical Fourier transform to calculate a spectrum from an interferogram appears to be Fellgett [7] in 1949, and this date heralds the beginning of interferometry as a viable spectroscopic technique. Fellgett was principally interested in astronomical observations where the intensity of the radiation being measured is extremely low. He was investigating methods of *multiplexing* the spectral information so that less radiation was lost on the jaws of the slits of a monochromator and realized that the use of a Michelson interferometer could yield a huge reduction in the time taken to measure the spectra of weak sources. The fact that data from all of the spectral frequencies are measured simultaneously during the complete measurement accounts for the principal advantage of interferometry. The reduction in measurement time which results from measuring all of the radiation during all of the measurement is usually known as *Fellgett's advantage* in honor of the man who first realized its potential. If longer times are taken for the measurement, spectra of greater signal-to-noise ratio are measured in proportion to the square root of the measurement time.

One other basic advantage in the use of the Michelson interferometer for spectroscopy is the achievement of a greater *throughput* of radiation than is possible with a monochromator measuring the same spectrum. "Throughput" is usually defined as the product of the area and solid angle of the beam from the source accepted by the instrument, and is alternatively known as "light grasp," "luminosity," and "étendue." The increased spectral signal-to-noise ratio resulting from the increased signal at the detector is often called *Jacquinot's advantage* after the French scientist who first pointed out the benefits that could accrue by combining the throughput advantage (which was already known for other interferometric methods such as the Fabry-Perot interferometer) with the multiplex advantage of the Michelson interferometer [8].

Unlike Fellgett's advantage, Jacquinot's advantage is not of such universal importance for chemical spectroscopy. In many applications the throughput allowed by the size or geometry of the sample or source may be much smaller than the maximum throughput allowed theoretically, and the throughput of most commercial infrared Fourier transform spectrometers is usually little different from that of grating spectrometers designed for the same type of measurement.

For visible spectroscopy, however, the situation is reversed. Whereas the noise level of most infrared detectors is independent of the signal, the noise level from photomultiplier tubes, which are commonly used for the measurement of visible radiation, increases with the signal. This effect obviously offsets Fellgett's advantage when sources of continuous radiation are being measured, since by allowing all the radiation to reach the photomultiplier simultaneously, the noise in the measured signal is increased. On the other hand, Jacquinot's advantage still gives substantial benefits for the measurement of the spectra of large sources of visible radiation, so that the use of interferometers for measuring visible spectra has been suggested for certain special applications.

In the 15 years following the work of Fellgett, interferometric spectroscopy was largely used only by those spectroscopists who were unable to obtain their results using a grating spectrometer because of measurement time considerations. Fourier transform spectroscopy was not attractive to the chemist mainly because of the inconvenience of the computation of the spectrum from the interferogram. When the number of data points of the interferogram that had to be collected and transformed was greater than 1000, the Fourier transform would take several minutes and when high resolution spectra were being computed from interferograms with over 10,000 data points, the computation time was several hours. The computing time in this case would offset any advantage to the chemist derived from the use of an interferometer for measuring his spectrum unless the energy being detected was extremely weak.

The type of weak sources that were being studied in the early days of Fourier transform spectroscopy included distant stars, and it is interesting to note that not only Fellgett but also many of the other leaders of Fourier transform spectroscopy have been astronomers. Among these are Pierre and Janine Connes, who have designed interferometers capable of yielding extremely high resolution spectra and developed the mathematical techniques necessary for the transformation of the million or so data points that they generate. Mertz, to whom the theory behind many of the more important advances in computational procedures can be ascribed and who designed the first rapid-scanning interferometers, is also an astronomical spectroscopist.

The first chemical applications of Fourier transform spectroscopy took place in the traditionally energy-starved far-infrared region of the spectrum. The use of interferometers for far-infrared spectroscopy has several benefits. First, since most of the interferograms that were measured had only a limited number of data points, the time required for the computations was relatively small. Also, the mechanical tolerance on the drive for the moving mirror of the Michelson interferometers used for measuring medium resolution far-infrared spectra is far less stringent than for the type of drive required for chemical spectroscopy in the mid-infrared, and orders of magnitude less than that required for Connes ultrahigh resolution near-infrared interfero-meter. Thus, interferometers for far-infrared spectroscopy could be easily and cheaply built commercially.

The success of interferometry in the far infrared caused many chemists to foresee applications for mid-infrared Fourier transform spectroscopy. Surprisingly enough, the first interferometers to be successfully used for mid-infrared spectroscopy were not modifications of an existing far-infrared interferometer but rather were derived from small rapid-scanning interferom-eters developed by Mertz [9] for ultrahigh sensitivity applications such as remote sensing and astronomy where high resolution was not required. The interferograms from this instrument were short so that data reduction did not present an insuperable problem, but the first rapid-scanning interferometers did not achieve wide acceptance by the chemical community due mainly to the low resolution of the spectra.

A real breakthrough for mid-infrared Fourier transform spectroscopy occurred in 1966, when the so-called Cooley-Tukey [10] fast Fourier trans-form algorithm (the FFT) was first applied to interferometry [11]. Computa-tions which prior to the use of this algorithm took hours to complete take merely minutes using the FFT, and low resolution spectra can be computed in a few seconds. Thus, it became feasible for spectroscopists who previously took hours over the measurement of a spectrum with many resolution elements using a monochromator to use an interferometer for the same measurement and show a considerable time saving.

At about the same time, the technology of solid-state devices and integrated circuits had advanced to the point that small general-purpose computers were being developed. The period between 1968 and the present day has signified not only the development of medium resolution interferometers suitable for chemical spectroscopy but also the interface of these instruments to dedicated minicomputers so that spectra can be plotted only seconds after *starting* the measurement of an interferogram. Spectra from these automated instruments are comparable in specifications to those produced on a grating spectrometer: the resolution is slightly higher while the frequency range is slightly less. However, the measurement time for a spectrum measured in-terferometrically is much less than that for the same spectrum measured by a grating spectrometer.

Infrared Fourier transform spectrometers are now available commercially as a routine tool for the analytical chemist and chemical spectroscopist. Al-though not every analytical laboratory needs a Fourier transform spectrom-eter, there are certainly many problems for which the use of this technique can provide an answer. It is hoped that this book will provide the chemical spectroscopist with adequate background material for an understanding of interferometry and reasons for its occasional failures.

The book is divided into two parts. Part A provides an introduction to the basic theory of Fourier transform spectroscopy and its relationship to instrumental design. The physics and mathematics required for comprehending the material covered in this section have been kept at a fairly low level and should be understandable by any reader with a first degree in chemistry. Occasionally, simplifications have been made in the derivation of formulas for the sake of clarity, but all the equations that are derived may be used in practice to a very good degree of approximation. Part B covers most of the more important published applications of Fourier transform spectroscopy in various branches of chemistry and several results in related areas have also been described.

REFERENCES

1. A. A. Michelson, *Phil. Mag.* Ser. 5, **31**, 256 (1891).
2. A. A. Michelson, *Phil. Mag.* Ser. 5, **34**, 280 (1892).
3. A. A. Michelson, *Light Waves and Their Uses*, University of Chicago Press, Chicago 1902 (reprinted, Pheonix Science Series, University of Chicago Press, 1962).
4. Lord Rayleigh, *Phil. Mag.* Ser. 5, **34**, 407 (1892).
5. H. Rubens and H. Hollnagel, *Phil. Mag.* Ser. 6, **19**, 761 (1910).
6. H. Rubens and R. W. Wood, *Phil. Mag.* Ser. 6, **21**, 249 (1911).
7. P. Fellgett, Aspen Int. Conf. on Fourier Spect., 1970 (G. A. Vanasse, A. T. Stair, and D. J. Baker, Eds.), AFCRL–71–0019, p. 139.
8. P. Jacquinot, 17^e Congres du GAMS, Paris, 1954.
9. L. Mertz, *Astron. J.*, **70**, 548 (1965).
10. J. W. Cooley and J. W. Tukey, *Math. Comput.*, **19**, 297 (1965).
11. M. L. Forman, *J. Opt. Soc. Am.*, **56**, 978 (1966).

PART A

Theory and Instrumentation

1

THE MICHELSON INTERFEROMETER

I. INTRODUCTION

The design of most interferometers being used for infrared spectroscopy today is based on that of the first interferometer originally designed by Michelson in 1891 [1, 2]. Many other two-beam interferometers have been designed subsequently which may sometimes be more useful for certain specific applications than the Michelson interferometer; however, the theory behind all scanning interferometers is similar, and the general theory of interferometry is most readily comprehended by first acquiring an under-standing of the way in which a basic Michelson interferometer can be used for the measurement of infrared spectra.

The Michelson interferometer is a device that can split a beam of radiation into two paths and then recombine them so that the intensity variations of the exit beam can be measured by a detector as a function of path difference. The simplest form of the interferometer is shown in Fig. 1.1. It consists of two mutually perpendicular plane mirrors, one of which can move along the axis shown. The movable mirror is either moved at a constant velocity or is held at equidistant points for fixed short time periods and rapidly stepped between these points. Between the fixed mirror and the movable mirror is a *beamsplitter*, where a beam of radiation from an external source can be

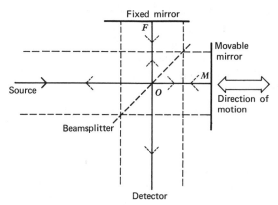

Fig. 1.1. Schematic diagram of a Michelson interfer-ometer.

9

partially reflected to the fixed mirror (at point F) and partially transmitted to the movable mirror (at point M). After each beam has been reflected back to the beamsplitter, they are again partially reflected and partially transmitted. Thus, a portion of the beams which have traveled in the path to both the fixed and movable mirrors reach the detector, while portions of each beam also travel back toward the source.

The beam that returns to the source is only occasionally of interest for spectroscopy and only the output beam passing in the direction perpendicular to that of the input beam is usually measured. However, it is important to remember that both of the output beams contain equivalent information. The main reason for measuring only one of the output beams is the difficulty of separating the second output beam from the input beam if the beamsplitter is at 45° to each mirror. In some measurements, both output beams are measured using two detectors; in others, separate input beams can be passed into each arm of the interferometer and the resultant signal measured using one or two detectors. These measurements are generally classified under the heading of "dual-beam interferometry" and are described in Chapter 7.

II. MONOCHROMATIC LIGHT SOURCES

To best understand the processes occurring in a Michelson interferometer, let us first consider an idealized situation. Consider the case of a source of monochromatic radiation producing an infinitely narrow, perfectly collimated beam. Let the wavelength of the radiation be λ cm, and its frequency be $\bar{v}$ cm^{-1}. For this example, suppose that the beamsplitter is a nonabsorbing film whose reflectance is 50% and whose transmittance is 50%. We first determine the intensity of the beam at the detector when the movable mirror is held stationary at different positions.

The path difference between the beams traveling to the fixed and movable mirrors is 2(OM − OF), and this optical path difference is called the *retardation*; it is usually given the symbol δ. Since δ is the same for all parallel input beams, such as the two broken lines shown in Fig. 1.1, we can now relax our criterion for an infinitely narrow input beam, but it should still remain collimated.

When the fixed and movable mirrors are equidistant from the beamsplitter (zero retardation), the two beams are perfectly in-phase on recombination at the beamsplitter (Fig. 1.2a). At this point the beams interfere *constructively* so that the intensity of the beam passing to the detector is the sum of the intensities of the beams passing to the fixed and movable mirrors. Therefore, all the light from the source reaches the detector at this point and none returns to the source.

If the movable mirror is displaced a distance $\lambda/4$ cm, the retardation is now $\lambda/2$ cm. Thus the path-lengths to and from the fixed and movable mirrors are exactly one half wavelength different so that they are out-of-phase and interfere *destructively* (Fig. 1.2*b*). At this point, all the light returns to the source and none passes to the detector. A further displacement of the movable mirror by $\lambda/4$ cm makes the total retardation λ cm. The two beams are once more in-phase on recombination at the beamsplitter and a condition of constructive interference again occurs, (Fig. 1.2*c*). For monochromatic radiation, there is no way of determining whether a particular point at which a

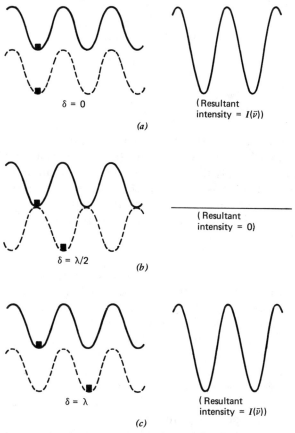

$\delta = 0$

(Resultant intensity = $I(\bar{\nu})$)

(a)

$\delta = \lambda/2$

(Resultant intensity = 0)

(b)

$\delta = \lambda$

(Resultant intensity = $I(\bar{\nu})$)

(c)

Fig. 1.2.   Diagram showing how the relative phases of the beams traveling to the fixed mirror (solid line) and movable mirror (broken line) change with the retardation, δ. The marker on each wave denotes light which left the source at the same time. The sum of the amplitudes of each wave gives the magnitude of the signal at the detector. Note that a maximum signal is measured when $\delta = n\lambda$ and that for monochromatic radiation there is no way to determine whether a given maximum is due to a retardation of zero or $n\lambda$.

signal maximum is measured corresponds to zero retardation or a retardation equal to an integral number of wavelengths.

If the signal at the detector is measured as the mirror is moved at constant velocity, it will be seen to vary sinusoidally, a maximum being measured each time that the retardation is an integral multiple of λ. The signal at the detector measured as a function of retardation is given the symbol $I'(\delta)$. The intensity of the signal at any point where $\delta = n\lambda$ (where n is an integer) is equal to the intensity of the source $I(\bar{v})$. At other values of δ, the magnitude of the signal at the detector is given by

$$I'(\delta) = 0.5\ I(\bar{v}) \left\{ 1 + \cos 2\pi \frac{\delta}{\lambda} \right\}. \tag{1.1}$$

Since $$\bar{v} = (\lambda)^{-1}, \tag{1.2}$$

it can be easily seen that

$$I'(\delta) = 0.5\ I(\bar{v}) \left\{ 1 + \cos 2\pi\bar{v}\delta \right\}. \tag{1.3}$$

It can be seen that this expression is composed of a constant (d.c.) component equal to $0.5\ I(\bar{v})$, and a modulated (a.c.) component equal to $0.5\ I(\bar{v})$ $\cos 2\pi\bar{v}\delta$. In practice, only the a.c. component is important in spectroscopic measurements, and it is this modulated component that is generally referred to as the *interferogram*, $I(\delta)$. In summary, the interferogram from a monochromatic source measured with an ideal interferometer is given by the equation

$$I(\delta) = 0.5\ I(\bar{v}) \cos 2\pi\bar{v}\delta. \tag{1.4}$$

In practice several factors affect the magnitude of the signal measured at the detector. First, it is practically impossible to find a beamsplitter having the ideal characteristics of 50% reflectance and 50% transmittance. Although this ideal situation can be approached fairly closely in practice, the non-ideality of the beamsplitter must be allowed for in Eq. 1.4 by multiplying $I(\bar{v})$ by a frequency-dependent factor less than unity representing the relative beamsplitter efficiency (see Chapter 4, section II). Second, most infrared detectors do not have a uniform response to all frequencies and for certain types of amplifiers, the amplification is also strongly frequency-dependent. It will be seen in Chapter 2 that the amplifier usually contains filter circuits to eliminate the signals from high and low frequency radiation reaching the detector; it is these filters which cause the amplifier to have a frequency-dependent response of the type illustrated later (Fig. 2.4).

Thus, in practice, the amplitude of the interferogram as observed after detection and amplification is proportional not only to the intensity of the source but also to the beamsplitter efficiency, detector response, and amplifier characteristics. Of these factors, only $I(\bar{v})$ varies from one measurement to the next for a given system configuration while all the other factors remain constant. Therefore Eq. 1.4 may be modified by a single frequency-dependent correction factor, $H(\bar{\nu})$, to give

$$I(\delta) = 0.5\ H(\bar{\nu})I(\bar{\nu})\cos 2\pi\bar{\nu}\delta. \qquad (1.5)$$

Putting $0.5\ H(\bar{\nu})I(\bar{\nu})$ equal to $B(\bar{\nu})$, it is seen that the simplest equation representing the interferogram is

$$I(\bar{\nu}) = B(\bar{\nu})\cos 2\pi\bar{\nu}\delta, \qquad (1.6)$$

and this is the form that will be most generally used throughout this book. The parameter $B(\bar{\nu})$ gives the intensity of the source at a frequency $\bar{\nu}$ cm^{-1} as modified by the instrumental characteristics.

When the movable mirror is being scanned at a constant velocity, V cm/sec, it becomes important to know the way in which the interferogram varies as a function of time, $I(t)$, rather than as a function of retardation, $I(\delta)$. Since the retardation is given by

$$\delta = 2Vt \text{ cm}, \qquad (1.7)$$

it is seen from Eq. 1.6 that

$$I(t) = B(f_{\bar{\nu}})\cos 2\pi\bar{\nu}\cdot 2Vt. \qquad (1.8)$$

The dimension used to reference the value of the interferogram (retardation in centimeters, or time in seconds) must always be the inverse of that used to reference the amplitude of the signal (optical frequency in cm^{-1}, or audio-frequency in Hertz, respectively). Thus, in Eq. 1.8, the amplitude of the cosine wave is given as $B(f_{\bar{\nu}})$, where $f_{\bar{\nu}}$ is the audio-frequency of the wave in Hertz.

For any cosine wave of frequency f Hz, the amplitude of the signal after a time t seconds is given by the equation:

$$A(t) = A_o \cos 2\pi\ ft, \qquad (1.9)$$

where A_o is the maximum amplitude of the wave. A comparison of Eqs. 1.8 and 1.9 shows that the audio-frequency, $f_{\bar{\nu}}$ Hz, of the interferogram $I(t)$ corresponds to a frequency of the radiation being emitted from the source, $\bar{\nu}$ cm^{-1}, through the relationship

$$f_{\bar{\nu}} = 2V\bar{\nu} \text{ Hz}. \qquad (1.10)$$

Mathematically, $I(\delta)$ is said to be the *cosine Fourier transform* of $B(\bar{\nu})$, while $I(t)$ is the cosine Fourier transform of $B(f_{\bar{\nu}})$. The spectrum is therefore determined from the interferogram by taking the Fourier transform of $I(\delta)$ or $I(t)$, which accounts for the name given to this spectroscopic technique— *Fourier transform spectroscopy.*

In the particular case where the spectrum of a source of monochromatic radiation is to be determined, performing the Fourier transform of a measured interferogram is a trivial operation. However, if the source is emitting discrete spectral lines of several frequencies or continuous radiation, the interferogram is more complex and a digital computer is usually required to perform the transform.

III. POLYCHROMATIC SOURCES

When more than one frequency is emitted by the source, the measured interferogram is the resultant of the interferograms corresponding to each frequency. For line sources with simple spectra, interferograms are often found which repeat themselves at regular intervals of retardation. Some simple spectra and their interferograms are shown in Fig. 1.3.

The upper curve in Fig. 1.3 represents the case when two closely spaced lines are examined. It is interesting to note that this was the situation that

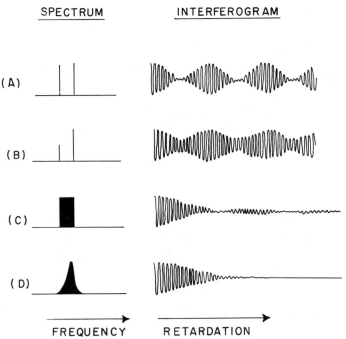

Fig. 1.3. Simple spectra and their interferograms: for (a) and (b), the smaller the separation of the spectral lines, the lower is the beat frequency. For (c) and (d), the smaller the width of the spectral band, the greater is the retardation at which the modulation of the interferogram becomes small.

occurred when Michelson examined the red Balmer line in the hydrogen spectrum [3]. Although he was not able to resolve the high frequency wave, Michelson observed the envelope, or visibility curve, of the interferogram. From this visibility curve, he concluded that the Balmer line is actually a doublet; we now know that this conclusion was correct, the separation of the two lines being 0.014 nm.

When the source is a continuum, the interferogram must be represented by an integral:

$$I(\delta) = \int_{o}^{+\infty} B(\bar{v}) \cos 2\pi\bar{v}\delta \cdot d\bar{v} \tag{1.11}$$

which is one-half of a cosine Fourier transform pair, the other being

$$B(\bar{v}) = \int_{-\infty}^{+\infty} I(\delta) \cos 2\pi\bar{v}\delta \cdot d\delta. \tag{1.12}$$

It may be noted that $I(\delta)$ is an even function, that is the interferograms to the left and right of the zero retardation point are equivalent. Thus, Eq. 1.12 may be rewritten as

$$B(\bar{v}) = 2 \int_{o}^{+\infty} I(\delta) \cos 2\pi\bar{v}\delta \cdot d\delta. \tag{1.13}$$

Equation 1.11 shows that *in theory* one could measure the complete spectrum from 0 to $+\infty$ cm^{-1} at infinitely high resolution, since $\delta\bar{v} \to 0$. However Eq. 1.13 shows that in order to achieve this, we would have to scan the moving mirror of the interferometer an infinitely long distance, with δ varying between 0 and $+\infty$ cm. Therefore, if the Fourier transform were to be performed using digital methods, the interferogram would have to be digitized at infinitesimally small intervals of retardation. *In practice*, it is found necessary to digitize the signal at finite intervals and this restriction has the effect of only allowing a certain frequency range to be measured; this problem will be discussed in more detail in Chapter 2. It is similarly apparent that the interferogram cannot, in practice, be measured to a retardation of $+\infty$ cm; the effect of measuring the signal over a limited retardation is to cause the spectrum to have a finite resolution, and this effect is described in the next section.

IV. FINITE RESOLUTION

It is fairly simple to illustrate how the resolution of a spectrum measured interferometrically depends on the maximum retardation of the scan. As an example, consider the case of a spectrum consisting of a doublet, both components of which have equal intensity. Figure 1.4a shows the spectrum, and Fig. 1.4b shows the interferogram from each line. Figure 1.4c shows the

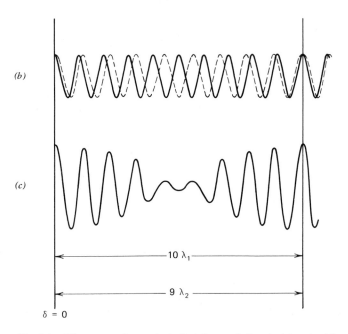

Fig. 1.4. Diagram to demonstrate that the resolution is determined by the retardation of the interferometer: (a) an emission spectrum with a doublet at frequencies $\bar{\nu}_1$ and $\bar{\nu}_2$, such that the separation of the lines, $\Delta\bar{\nu}$, is equal to $0.1\bar{\nu}_1$; (b) the interferograms from the individual lines $\bar{\nu}_1$ (solid line) and $\bar{\nu}_2$ (broken line); (c) the resultant of the two interferograms above, showing that they become in-phase at a retardation $(0.1\bar{\nu}_1)^{-1}$ cm, that is, $\delta = (\Delta\bar{\nu})^{-1}$.

resultant of these curves. This case is the same as the upper curve in Fig. 1.3.

If the doublet has a separation of $\Delta\bar{\nu}(=\bar{\nu}_1 - \bar{\nu}_2)$ cm^{-1}, the two cosine waves in Fig. 1.4b become out-of-phase after a retardation of $1/2\,(\Delta\bar{\nu})^{-1}$ cm and are once more back in-phase after a retardation of $(\Delta\bar{\nu})^{-1}$ cm. Thus, to go through one complete period of the beat frequency, a retardation of $(\Delta\bar{\nu})^{-1}$ cm is required. The narrower the separation of the doublet, the greater must be the retardation before the cosine waves become in-phase. It is therefore apparent that the spectral resolution depends on the maximum retardation of the interferometer. Intuitively it might be concluded that the two lines would just be resolved in a spectrum computed from this interferogram if the retardation were increased to the point when the two waves became in-phase for the first time after zero retardation. At this point

$$\Delta = (\Delta\bar{v})^{-1} \text{ cm},$$

where Δ is the retardation at the end of the scan. Thus, if the maximum retardation of an interferometer is Δ_{max}, the best resolution that could be obtained using this interferometer, $(\Delta\bar{v})$, is given by

$$\Delta\bar{v} = (\Delta_{max})^{-1} \text{ cm}^{-1}. \tag{1.14}$$

This conclusion was arrived at somewhat intuitively, but the answer proves to be approximately correct. The next few paragraphs will show a more rigorous mathematical verification of this conclusion.

By restricting the maximum retardation of the interferogram to Δ cm, we are effectively multiplying the complete interferogram (between $\delta = -\infty$ and $\delta = +\infty$) by a truncation function, $D(\delta)$, which is unity between $\delta = -\Delta$ and $+\Delta$, and zero at all other points, that is:

$$
\begin{aligned}
D(\delta) &= 1 &\text{if} \quad &-\Delta < \delta < +\Delta \\
D(\delta) &= 0 &\text{if} \quad &\delta < -\Delta \\
& &\text{or} \quad &\delta > +\Delta.
\end{aligned}
\tag{1.15}
$$

In view of the shape of this function, $D(\delta)$ is often called a *boxcar* truncation function. The spectrum in this case is given by the equation:

$$B(\bar{v}) = \int_{-\infty}^{+\infty} I(\delta)\, D(\delta) \cos 2\pi\bar{v}\delta \cdot d\delta. \tag{1.16}$$

Mathematically for any two functions, multiplication and convolution are Fourier transform pairs [4]. Thus the spectrum is given by the convolution of the Fourier transform of the complete interferogram (for $-\infty < \delta < +\infty$) with the Fourier transform of the boxcar truncation function, $D(\delta)$.

For an infinitely long cosine wave interferogram, the spectrum is known to be an infinitesimally narrow spike, while the cosine Fourier transform of $D(\delta)$ is a function $f(\bar{v})$ given by:

$$
\begin{aligned}
f(\bar{v}) &= 2\Delta \cdot \frac{\sin (2\pi\bar{v}\Delta)}{2\pi\bar{v}\Delta} \\
&\equiv 2\Delta \cdot \text{sinc } (2\pi\bar{v}\Delta).
\end{aligned}
\tag{1.17}
$$

This function is centered about $\bar{v} = 0$, and is shown in Fig. 1.5a. The function sinc x intersects the x axis at $x = n\pi$, where $n = 1, 2, 3, \ldots$ For the present example, intersection occurs when $2\bar{v}\Delta = n$. The frequency of the first intersection, at $n = 1$, is therefore given by

$$\bar{v} = (2\Delta)^{-1} \text{ cm}^{-1}. \tag{1.18}$$

When the sinc function from the transform of the boxcar truncation function is convolved with a single spectral line of frequency $\bar{v}_1$, the resultant curve has the formula

$$B(v) = 2B(\bar{v}_1) \text{ sinc } 2\pi(\bar{v}_1 - \bar{v})\Delta. \tag{1.19}$$

This curve is shown in Fig. 1.5b and represents the appearance of a single sharp line measured interferometrically at a resolution considerably lower than the half-width of the line. The function $f(\bar{v})$ has been variously called the *instrument line shape* (ILS) function, the instrument function, and the apparatus function.

Since the curve intersects the frequency axis at $(2\Delta)^{-1}$ cm^{-1} either side of $\bar{v}_1$, it can be seen that two lines separated by twice this amount, that is, by Δ^{-1} cm^{-1}, would be completely resolved, and the result derived earlier by intuition is indeed correct. In fact, since two lines separated by Δ^{-1} cm^{-1} are completely resolved, the practical resolution is somewhat better than this value.

Several criteria have been used to define the resolution of spectrometers of which the most popular are the Rayleigh criterion and the full-width at half-height (FWHH) method. The Rayleigh criterion was originally used to define the resolution obtainable from a diffraction-limited grating spectrometer, the ILS of which may be represented by a function of the form sinc^2 x. Under the Rayleigh criterion, two adjacent spectral lines of equal intensity with a sinc^2 x ILS are considered to be just resolved when the center of one line is at the same frequency as the first zero value of the ILS of the other. Under this condition, the resultant curve has a dip of approximately 20% of the maximum intensity in the spectrum, as shown in Fig. 1.6a. However, if the same criterion is applied to lines having a sinc x ILS, it is found that the two lines are *not* resolved, as shown in Fig. 1.6b.

The FWHH method has mainly been used for spectrometers with a triangular slit function. Two triangularly shaped lines of equal intensity and half-width are not resolved until the spacing between the lines is greater than the full-width at half-height of either line. The FWHH of a line having the ILS given by Eq. 1.17 is $0.605/\Delta$ cm^{-1}, and it can be shown that two lines with sinc x line shapes are not resolved when they are separated by this amount. In practice a dip of approximately 20% is found when the two lines are separated by $0.73/\Delta$ cm^{-1}, as shown in Fig. 1.6c.

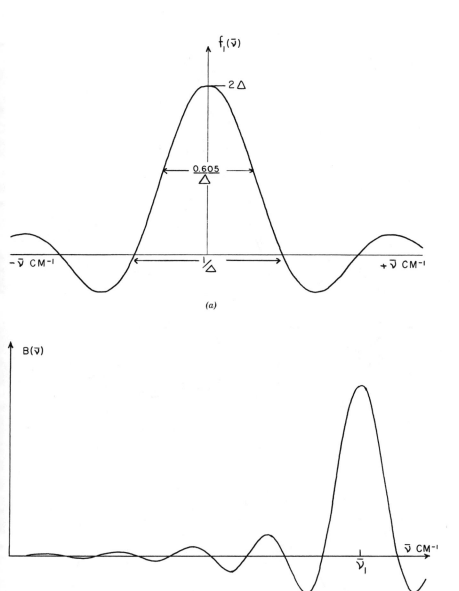

Fig. 1.5. (a) The function 2Δ sinc $2\pi\bar{\nu}\delta$, which is the result of performing the Fourier transform of the boxcar function $D(\delta)$; (b) the result of convolving this function with a single sharp spectral line at a frequency $\bar{\nu}_1$.

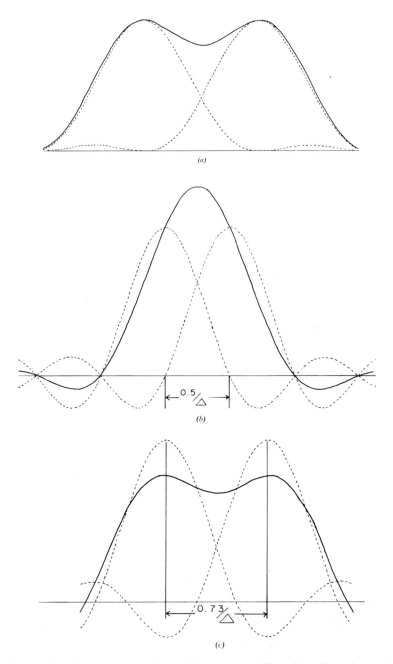

Fig. 1.6. (a) The Rayleigh criterion for resolving two spectral lines of equal intensity measured with a $\text{sinc}^2\ x$ ILS occurs when the maximum of one line coincides with the first zero value of the other; the resulting curve shows a dip of 20% between the two maxima; (b) when the same criterion is applied to two spectral lines with a $\text{sinc}\ x$ ILS, they are not resolved; (c) only when they are separated by $0.73/(\Delta)$ cm^{-1} does the dip in the resulting curve reach 20% of the maximum intensity.

The sinc x function is not a particularly useful line shape for infrared spectroscopy in view of the fairly large amplitude of the curve at frequencies well away from $\bar{v}_1$. The first minimum reaches below zero by an amount that is 22% of the height at $\bar{v}_1$. If a second weak line happened to be present in the spectrum at the frequency of this minimum, it would not be seen in the computed spectrum. One method of circumventing the problem of these secondary minima is through the process known as *apodization*.

V. APODIZATION

From the previous section we know that when a cosine wave interferogram is unweighted, the shape of the spectral line is the convolution of the transform of the cosine wave measured with an infinitely long retardation (i.e., the true spectrum) and a sinc function which is the transform of the boxcar truncation function, $D(\delta)$. If instead of using the boxcar function, we used a weighting function of the form:

and
$$
\begin{aligned}
A_1(\delta) &= 1 - |\delta/\Delta| \quad &&\text{for} \quad -\Delta < \delta < +\Delta \\
A_1(\delta) &= 0 \quad &&\text{for} \quad \delta < -\Delta \\
& &&\text{and} \quad \delta > +\Delta
\end{aligned} \tag{1.20}
$$

the true spectrum would be convolved with the Fourier transform of $A_1(\delta)$, and this function would therefore determine the ILS. The Fourier transform of $A_1(\delta)$ has the form:

$$
\begin{aligned}
f_1(\bar{v}) &= \Delta \cdot \frac{\sin^2 (\pi \bar{v} \, \Delta)}{(\pi \bar{v} \, \Delta)^2} \\
&\equiv \Delta \cdot \operatorname{sinc}^2 (\pi \bar{v} \, \Delta).
\end{aligned} \tag{1.21}
$$

This function is shown in Fig. 1.7, and it can be seen that the amplitude of the side-lobes has been considerably reduced from that of the side-lobes for the sinc function shown in Fig. 1.5a. Suppression of the magnitude of these oscillations is known as *apodization*,* and functions such as $A_1(\delta)$ which weight the interferogram for this purpose are known as *apodization functions*.

The specific function $A_1(\delta)$ is called a *triangular* apodization function and is the most common apodization function used in infrared Fourier transform spectroscopy, since it not only decreases the amplitude of the side-lobes (which all apodization functions do) but it also generates an ILS which is exactly the same as that of a diffraction-limited grating spectrometer. Thus, for lines separated by $1/\Delta$ cm^{-1}, a 20% dip would be found, as shown in Fig. 1.6a, while if they were separated by $2/\Delta$ cm^{-1}, the lines would be fully

---

* The word apodization refers to the suppression of the side-lobes (or feet) of the ILS; the word is apocryphally derived from the Greek, $\alpha \, \pi o \delta o s$ (without feet)!

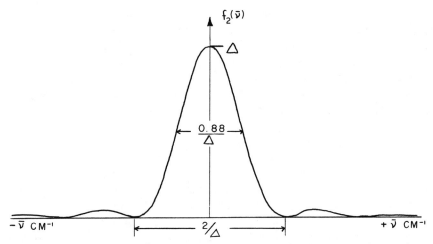

Fig. 1.7. The function Δ sinc^2 ($\pi\bar{\nu}\delta$), which results from performing the Fourier transform of the triangular function $A_1(\delta)$; note that although the frequency interval between the first zero values of this function is twice that of the corresponding sinc function shown in Fig. 1.5a, its FWHH is less than 1.5 times the FWHH of the sinc function.

resolved. The full-width at half-height for the function $f_1(\bar{\nu})$ is $0.88/\Delta$ cm^{-1}, and lines separated by this amount are just resolved; however, the dip is extremely small, being of the order of 1%.

In an absorption experiment, the measured interferogram is the sum of the interferogram of the source with no sample present and the interferogram due to the sample; since energy is being absorbed by the sample, these two interferograms are 180° out-of-phase. When the measured interferogram is triangularly apodized, the effect of apodization may therefore be considered separately for the source and sample interferogram. The interferogram due to a broad-band continuous source has the appearance of an intense spike at zero retardation about which can be seen modulations which rapidly die out to an unobservably low amplitude. On the other hand, the interferogram caused by a sample with very narrow absorption lines will show appreciable modulation even at large retardations. When the measured interferogram is triangularly apodized, the background spectrum due to the source is affected very little since its interferogram has very little information at high retardation, whereas the narrow absorption *will* be apodized and will show a sinc^2 function ILS.

Another commonly used apodization function has the shape of a trapezium (Fig. 1.8a). By adjusting the position of the point B, the resolution and the ILS can be easily altered. If B is placed at A, the function becomes identical to $A_1(\delta)$, whereas if it is placed at C, the function is identical with $D(\delta)$. If B is situated between A and C, the line shape will be intermediate between a sinc and a sinc^2 shape, so that the position of B can serve to provide a fine control on the resolution between the extremes given by $D(\delta)$ and $A_1(\delta)$. The instrument line shape can therefore be controlled by the exact nature of the apodization function applied, and many different apodization functions have been proposed for infrared spectroscopy. For example, a Gaussian apodization function of the form

$$A_2(\delta) = \exp\left(-a^2\delta^2\right)$$

(Fig. 1.8b) gives a Gaussian ILS[5], whereas a Lorentzian ILS can be generated using an apodization function of the form

$$A_3(\delta) = \exp\left(-a|\delta|\right)$$

(Fig. 1.8c). Another function that has also been used is a quarter-wave cosine function of the type

$$A_4(\delta) = \cos\left\{\frac{\pi}{2}\left|\frac{\delta}{\Delta}\right|\right\}$$

The shape of the ILS derived using the latter apodization function is rather similar to that obtained from the triangular apodization function, but in view of the increased weight given to the points close to the zero retardation point, the FWHH is somewhat narrower and the side-lobes are slightly more pronounced [5]. The exponential and quarter-wave cosine apodization functions have the disadvantage that they take longer to generate in a digital computer, and the most commonly used and the most useful apodization function for general purpose infrared spectroscopy is the triangular function shown in Eq. 1.20.

It is interesting at this point to consider the effect of multiplying the interferogram by other functions which, although they have been called apodization functions [6], do not apodize the interferogram. Consider a function which, rather than reducing the value of the interferogram at high retardation (the *wings* of the interferogram), reduces it about zero retardation, such as the function shown in Fig. 1.8d. In this case the high resolution information in the wings of the interferogram is given more weight so that the width of the ILS is actually decreased below that of the same interferogram transformed using the boxcar truncation function, $D(\delta)$. Although this technique does have the effect of enhancing the resolution of the spectrum, it has the disadvantage of producing a spectrum with such prominent side-lobes that it becomes very difficult to distinguish between real spectral features and instrumental artifacts.

VI. PHASE CORRECTION

Up to this point, it has been assumed that Eqs. 1.11 and 1.13 give an accurate representation of the interferogram; in practice, additional terms

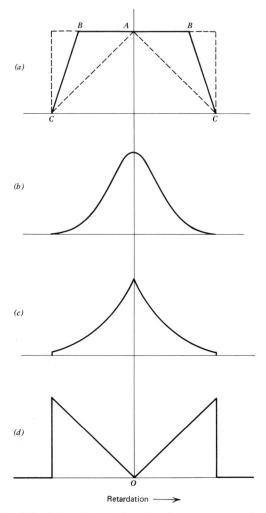

Fig. 1.8. Other apodization functions encountered in Fourier transform spectroscopy: (a) a programmable trapezium; if $B = C$, the function is a boxcar, and if $B = A$, the function is a triangle. If B is placed between these two extremes the ILS is intermediate between sinc x and sinc^2 x; (b) a Gaussian function, exp $(-a^2\delta^2)$; the Fourier transform of this function is another Gaussian function; (c) an exponential function, exp $(-a|\delta|)$; the Fourier transform of this function is a Lorentzian; (d) an "inverse triangular" function; the Fourier transform of this function has a slightly smaller half width than the sinc x function given using a standard boxcar truncation (thereby giving some resolution enhancement) but has larger side-lobes.

often have to be added to the phase angle, $2\pi\bar{v}\delta$, to describe the actual measured interferogram. Corrections to the phase angle may arise due to optical, electronic, or sampling effects; two common examples which lead to a change in the cosine term of these equations may be cited.

1. If we make use of the fact that the interferograms as represented in these equations are symmetrical about $\delta = 0$ and sample just one side of the interferogram, we must collect the first data point at exactly $\delta = 0$. If the first data point is actually sampled before the zero retardation point, at $\delta = -\varepsilon$, the interferogram takes the form

$$I(\delta) = \int_0^{+\infty} B(\bar{v}) \cos 2\pi\bar{v}(\delta - \varepsilon) \cdot d\bar{v}. \tag{1.22}$$

2. Electronic filters designed to remove high-frequency noise from the interferogram have the effect of putting a frequency dependent phase lag, $\theta_{\bar{v}}$, on each cosinusoidal component of the interferogram, and the resultant signal is given by

$$I(\delta) = \int_0^{+\infty} B(\bar{v}) \cos (2\pi\bar{v}\delta - \theta_{\bar{v}}) \cdot d\bar{v}. \tag{1.23}$$

Since any cosine wave, $\cos (\alpha + \beta)$ can be represented by:

$$\cos (\alpha + \beta) = \cos \alpha \cdot \cos \beta - \sin \alpha \cdot \sin \beta, \tag{1.24}$$

the addition of a constant to the phase angle, $2\pi\bar{v}\delta$, has the effect of adding sine components to the cosine wave interferogram.

The cosine Fourier transform of a truncated sine wave has the form shown in Fig. 1.9 so that, unless $\theta_{\bar{v}} = n\pi/2$ (where n is an integer), the ILS is intermediate between this function and the sinc function resulting from the cosine transform of a truncated cosine wave. The process of removing these sine components from an interferogram, or removing their effects from a spectrum, is known as *phase correction*.

Connes [7] has discussed the effect on the ILS due to an error in the choice of the zero retardation point. She defined a parameter called the phase displacement, which is the ratio of the sampling error, ε in Eq. 1.22, to the wavelength of the line being measured, λ. The way in which the ILS changes as the phase displacement varies between 0 (pure cosine wave) and 0.25 (pure sine wave) is reproduced in Fig. 1.10a. For even greater phase displacements, the spectrum can go completely negative (when $\varepsilon/\lambda = 0.5$) and return positive when the sampling error is exactly one wavelength. Connes has shown how the ILS varies as the phase displacement varies from -1 to $+1$, and this is shown in Fig. 1.10b. This is equivalent to varying the phase angle, $\theta_{\bar{v}}$, from -2π to $+2\pi$. From these diagrams, it can be seen that unless very accurate phase correction is carried out, gross photometric errors can result in the spectrum.

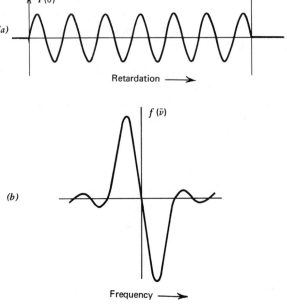

Fig. 1.9. If the cosine transform of the truncated *sine* wave (*a*) is performed, the resulting ILS has the shape shown in (*b*); this condition would be encountered if the first data point was sampled one quarter wavelength displaced from zero retardation ($\varepsilon/\lambda = 0.25$).

When a sampling error, ε, exists for a symmetrical interferogram, it may be estimated by sampling several data points about the maximum signal. A parabolic curve can be made to fit these points to fairly good accuracy, so that an estimate of ε can be made; the values of δ used in the transform can then be readily corrected for the sampling error.

Another way to solve the problem of not sampling the interferogram at the zero retardation point is to collect the data equally on both sides of $\delta = 0$, and perform the transform *double-sided*. For a double-sided transform, both the cosine transform, $T_{\cos}(\bar{v})$, and the sine transform, $T_{\sin}(\bar{v})$, are computed, and the *power spectrum*, $|\hat{S}(\bar{v})|$, is given by

$$|\hat{S}(\bar{v})| = \{[T_{\cos}(\bar{v})]^2 + [T_{\sin}(\bar{v})]^2\}^{\frac{1}{2}}. \tag{1.25}$$

The advantage of this technique over the single-sided transform, where data are only collected on one side of the zero retardation point, is that it requires no precise knowledge of the position of the zero retardation point. There are, however, several disadvantages to the method. First, it requires the moving mirror of the interferometer to be translated twice as far as in

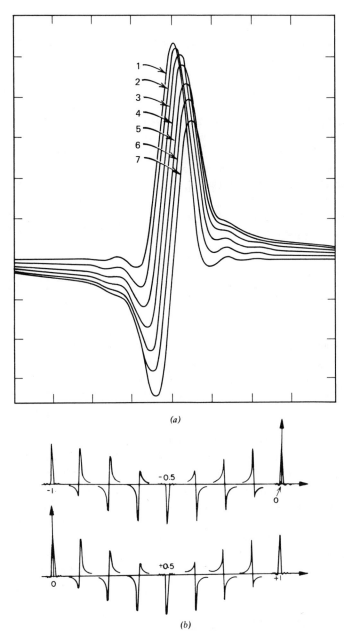

(a)

(b)

Fig. 1.10 (a) Variation of the ILS with phase displacement: 1, $\varepsilon/\lambda = 0$; 2, $\varepsilon/\lambda = 0.042$; 3, $\varepsilon/\lambda = 0.084$; 4, $\varepsilon/\lambda = 0.126$; 5, $\varepsilon/\lambda = 0.168$; 6, $\varepsilon/\lambda = 0.210$; 7, $\varepsilon/\lambda = 0.25$. For this figure, the sinusoid was apodized with the quarter-wave function, $A_4(\delta)$ described in Chapter 1, Section V. (b) When $\varepsilon/\lambda = 0.5$, the ILS is an inverted sinc function; this figure shows how the ILS changes as the phase displacement is varied from -1 to $+1$. (Reproduced from [7] by permission of the author.)

the single-sided method to achieve the same resolution. This in turn means that the measurement time is twice as long and the computation time is about four times longer. Another fundamental disadvantage is found in the fact that all the real spectral information *and the noise* is computed to have a positive value because the squares of the sine and cosine transforms are added before the square root is taken. Therefore, this method does not give a true representation of noise, which is, of course, randomly positive and negative. For a noisy spectrum, the height of the baseline is shifted above its true value, giving rise to an effective photometric error.

In the treatment of Connes [7], the interferogram was assumed to be symmetrical and the phase angle was only considered to be due to a sampling error, so that the phase angle, $\theta_{\bar{v}}$, in Eq. 1.23 was constant. In the more general case, $\theta_{\bar{v}}$ varies with frequency. If $\theta_{\bar{v}}$ varies linearly with frequency:

$$\theta_{\bar{v}} = 2\pi\kappa\bar{v}, \tag{1.26}$$

then,

$$I(\delta) = \int_0^{+\infty} B(v) \cos 2\pi v(\delta - \kappa) \cdot dv, \tag{1.27}$$

and the interferogram is symmetrical about the point where $\delta = \kappa$. The position where all frequencies add constructively is called the point of *stationary phase* [8], and it can be seen from Eq. 1.27 that this point is not necessarily equal to the zero retardation point.

When higher order terms are present in the phase spectrum, for example,

$$\theta_{\bar{v}} = A + B\bar{v} + C\bar{v}^2 + \cdots, \tag{1.28}$$

then there is no point of stationary phase and the interferogram is said to be *chirped*. For most interferograms, the amount of chirping is small; Fig. 1.11 shows an interferogram exhibiting a small amount of chirping.

More general phase correction routines have to be used when $\theta_{\bar{v}}$ varies with frequency; when the variation is slow, data collection is started a short distance to the left of zero retardation, at $-\Delta_1$, and continued until the desired resolution has been attained, that is, to a point such that $\delta = +\Delta_2$, where $(\Delta_2)^{-1}$ is the resolution desired. The short double-sided region of the interferogram from $-\Delta_1$ to $+\Delta_1$ is used to calculate the phase spectrum, $\theta_{\bar{v}}$, by computing the sine and cosine transforms and applying the relationship:

$$\theta_{\bar{v}} = \arctan\left\{\frac{T_{\sin}(\bar{v})}{T_{\cos}(\bar{v})}\right\}. \tag{1.29}$$

In the most commonly used method of phase correction, which was developed by Mertz [9], the amplitude spectrum is calculated with reference to the point of stationary phase or, in the case of slightly chirped inter-

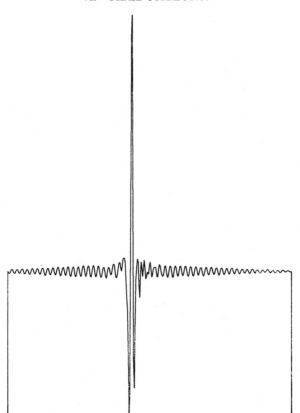

Fig. 1.11.   A slightly chirped interferogram; this interferogram is typical of the signal measured by rapid-scanning interferometers.

ferograms, to the largest data point in the array. The calculated amplitude spectrum is then multiplied at each frequency by the cosine of the difference between the measured phase angle and the reference phase angle to yield the true phase corrected spectrum.

Prior to the initial calculation of the amplitude spectrum, the original interferogram must be multiplied by a weighting function of the type shown in Fig. 1.12a, since if all the points in the interferogram were weighted equally, the region around the zero retardation point would have been counted twice with respect to the more distant fringes, giving rise to a source of photometric error. Narrow absorption bands give rise to modulations in the interferogram which do not die out until the retardation is large, whereas the modulations caused by a broad-band source occur mainly in the short,

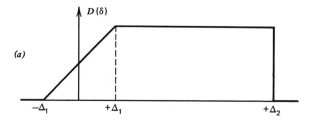

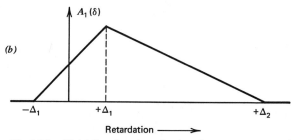

Fig. 1.12. Weighting functions for interferograms measured for phase correction by the Mertz method: (a) gives the equivalent of boxcar truncation; (b) gives the equivalent of triangular apodization.

double-sided region of the interferogram (Fig. 1.13). Thus, unless the interferogram is weighted in the manner shown in Fig. 1.12a, the intensities of sharp absorption bands would appear to be about half of their true value, as shown in Fig. 1.14. The ILS found using this type of weighting function is a sinc function, and if a $sinc^2$ function is desired for the ILS, an apodization function of the type shown in Fig. 1.12b must be applied.

Another method of phase correction has been described by Forman, Steel, and Vanasse [10], which involves convolving the *interferogram* with a phase error function calculated from the short, double-sided interferogram in order to symmetrize the interferogram, after which the cosine Fourier transform is computed. This method also yields good results, although the computation time is a little longer than the Mertz method, especially for long interferograms, since convolution is a longer procedure than multiplication.

The phase correction techniques of both Mertz [9] and Forman et al. [10] allow the interferometer to be used more efficiently than the double-sided mode, since a resolution corresponding to almost the full length of the scan is obtained, computing time is reduced, noise is randomly positive and negative (rather than all positive as in the double-sided technique), and the signal-to-noise ratio in the spectrum is just what would be expected from the interferogram.

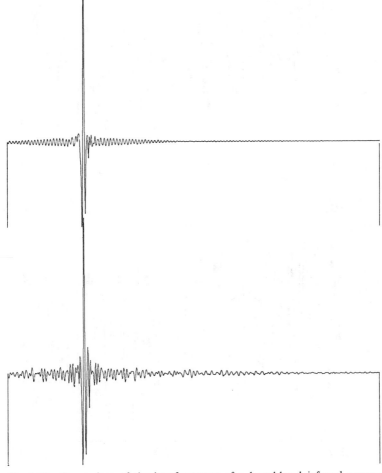

Fig. 1.13.   Comparison of the interferograms of a broad-band infrared source: (a) with no sample in the beam, and (b) with a 0.05 mm sheet of polystyrene placed between the interferometer and the detector. (Small modulations just apparent in the wings of the upper signal are caused by a small amount of atmospheric water vapor in the beam.)

VII.   EFFECT OF BEAM DIVERGENCE THROUGH THE INTERFEROMETER

In Section II it was assumed that the beam passing through the interferometer is perfectly collimated. In practice a perfectly collimated beam could only be generated from an infinitely small source, and in order to

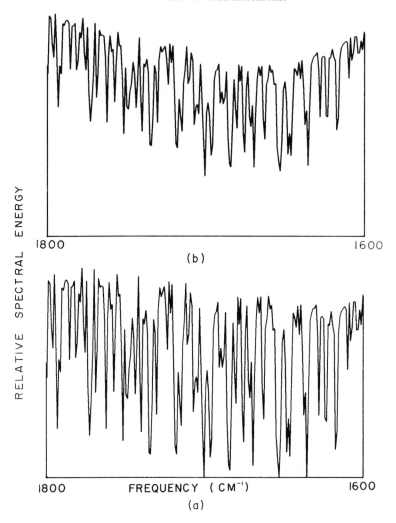

Fig. 1.14. The single-beam spectrum from 1800 to 1600 cm^{-1} of a broad-band IR source with a 1 m absorbing path of water vapor between the source and the detector, measured at 2 cm^{-1} resolution: (a) measured using an interferogram taken from -0.03 cm to $+0.5$ cm retardation, computed using the weighting function shown in Fig. 1.12a; (b) computed from the same interferogram with all data points given equal weight. It is seen that in (b) the intensity of each line is approximately half the intensity of the corresponding line in (a).

measure any signal at the detector a source of finite size must be used. In this section the limits on source size are discussed in order to determine the largest throughput of radiation that it is possible to use without degrading the spectrum in any way.

If the light from an extended monochromatic source is passed through a

Michelson interferometer, a circular fringe pattern is produced in the plane of the image of the source [11] (e.g., the detector position). This effect is shown schematically in Fig. 1.15 [8]. The diameter of the rings is a maximum at zero

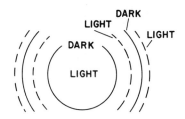

DARK
LIGHT
DARK
LIGHT
LIGHT

Fig. 1.15. (a) Ring pattern in the plane of the detector for a Michelson interferometer illuminated with monochromatic light with a finite solid angle; (b) intensity variations across the pattern above. (Reproduced from [11] by permission of the author.)

retardation and shrinks as the retardation is increased. As the retardation is increased, the intensity of the signal at any point in this plane changes sinusoidally. Unless the aperture at a plane of the source image is limited to include only one fringe of the interferogram, there will be no overall change of intensity with retardation and no interferogram would be able to be recorded.

This can be viewed in another way so that quantitative relationships can be found. Consider the effect of a noncollimated beam of monochromatic light of wavelength λ cm, passing through the interferometer with a divergence half-angle α. At zero retardation, the path difference between the central ray passing to the fixed and movable mirrors is zero, and there is also no path difference for the extreme rays. Thus, there will be constructive interference for both beams. Let the movable mirror now be moved a distance l (Fig. 1.16). The increase in retardation for the central ray is $2l$, while for the extreme ray it is $2l/\cos \alpha$. Therefore, a path difference, x, has been generated between the central and the extreme ray, where

$$x = \frac{2l}{\cos \alpha} - 2l$$

$$= 2l \left\{ \frac{1 - \cos \alpha}{\cos \alpha} \right\}. \tag{1.30}$$

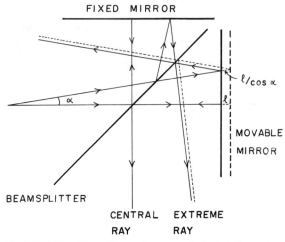

Fig. 1.16. The effect of beam divergence in the interferometer: the path difference for the central ray is 2*l* when the mirror is moved a distance 1, while that for the extreme ray is 2*l*/cos α.

Since

$$\cos \alpha = 1 - \frac{\alpha^2}{2!} + \frac{\alpha^4}{4!} - \frac{\alpha^6}{6!} + \cdots ,$$

if α is small

$$1 - \cos \alpha \sim \frac{\alpha^2}{2};$$

therefore, since cos α ~ 1

$$x = 2l \cdot \frac{\alpha^2}{2} = l\alpha^2. \tag{1.31}$$

As *l* increases, the extreme ray will be out-of-phase with the central ray for the first time when:

$$x = \frac{\lambda}{2}.$$

At this point the fringe contrast at the detector completely disappears, and any further increase in *l* will add no further information to the interferogram (Fig. 1.17). This retardation, 2*l*, gives the highest resolution, $\Delta \bar{\nu}$, achievable with this half-angle, α, for a wavelength λ cm or frequency $\bar{\nu}(=\lambda^{-1})$ cm^{-1}. (By Eq. 1.14, $(2l)^{-1} = \Delta \nu$). Fringe contrast would obviously have been lost at a shorter retardation if $\bar{\nu}$ were increased. Therefore, if a resolution of $\Delta \bar{\nu}$ is to be achieved at all frequencies in a spectrum whose highest frequency is $\bar{\nu}_{max}$, the greatest beam half-angle, α_{max}, that can be passed through the inter-

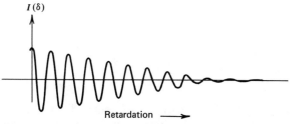

Fig. 1.17.   The type of interferogram that would be measured from a highly divergent monochromatic beam; note the similarity between this interferogram and the one shown in Fig. 1.3*d*.

ferometer is given by

$$l \, \alpha^2 {}_{max} \left\{ = \frac{\alpha^2}{2(\Delta \bar{v})} \right\} = \frac{1}{2 \bar{v}_{max}}.$$

Therefore

$$\alpha_{max} = \left\{ \frac{\Delta \bar{v}}{\bar{v}_{max}} \right\}^{\frac{1}{2}}. \tag{1.32}$$

Besides setting a limit on the resolution achievable, beam divergence also has the effect of shifting the frequency of a computed spectral line from its true value. Consider the interferograms due to the central and extreme rays from a monochromatic source. The wavelength of the interferogram for the extreme ray is longer than that of the central ray (Fig. 1.16). For a particular divergence half-angle, α, the path difference between the central and extreme rays at a retardation Δ is given by Eq. 1.31 as

$$x = \frac{\alpha^2 \Delta}{2}.$$

At this retardation there are n maxima in the (cosine) interferogram for the central ray, where

$$\Delta = n\lambda = \frac{n}{\bar{v}}. \tag{1.33}$$

For the extreme ray, there is an increased retardation, $(\Delta + x)$, and the effective wavelength of this ray is therefore changed to a value λ', where

$$\Delta + x = n\lambda' = \frac{n}{\bar{v}'}. \tag{1.34}$$

Combining Eqs. 1.33 and 1.34, and substituting for x from (1.32), we obtain

$$\frac{\bar{v}'}{\bar{v}} = \frac{\Delta}{\Delta + x} = \frac{1}{1 + \alpha^2/2}. \tag{1.35}$$

If α is small

$$\frac{\bar{v}'}{\bar{v}} = 1 - \frac{\alpha^2}{2}$$

so that

$$\bar{v}' = \bar{v}\left\{1 - \frac{\alpha^2}{2}\right\}. \tag{1.36}$$

The actual frequency of the line which is calculated from the interferogram, $\bar{v}''$, is approximately equal to the mean of the frequencies of the central and extreme rays, so that

$$\bar{v}'' = \frac{\bar{v} + \bar{v}'}{2} = \bar{v}\left\{1 - \frac{\alpha^2}{4}\right\}. \tag{1.37}$$

In practice the correction factor, $\alpha^2/4$, is very small, but when high accuracy is required it can be quite significant.

VIII. EFFECT OF MIRROR MISALIGNMENT

Two effects dependent on the alignment of the mirrors in a Michelson interferometer can affect the quality of a spectrum. The first depends on the alignment of the fixed mirror relative to the moving mirror, and the second depends on how accurately the plane of the moving mirror is maintained during the scan. These effects will be discussed in this section and in the following one, respectively.

If the moving mirror is held at a different angle than the fixed mirror relative to the plane of beamsplitter, then the image of the beam to the moving mirror will hit the plane of the detector at a different position than the beam which traveled to the fixed mirror. Consider the concentric rings forming the image of an extended source at the detector (Fig. 1.15). If the images from the fixed and moving mirrors are not centered at the same point on the detector, fringe contrast can be drastically reduced. Since the diameter of any ring is dependent on the frequency of the radiation, high frequencies from a polychromatic source will be affected more than low frequencies. Thus, a small reduction in the amplitude of the low-frequency waves in the interferogram may be noted when the interferometer goes out of alignment, while the amplitude of the high-frequency waves can be seriously reduced.

This effect is shown in Fig. 1.18; spectra are shown where the two mirrors in the interferometer are (a) well aligned, (b) in fair alignment, and (c) badly aligned. It can be seen that as the interferometer goes out of alignment, the frequency at which the greatest intensity in the spectrum occurs is reduced, so that when the interferometer goes badly out of alignment all the information at high frequency is lost.

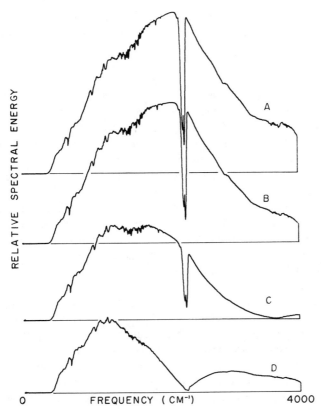

Fig. 1.18. Single-beam spectra taken when the interferometer is in (a) good alignment, (b) fairly good alignment, (c) poor alignment, and (d) very poor alignment. The energy at high frequencies for (b) is only slightly lower than for (a), but in (c) it is seen to be much lower. For very poorly aligned interferometers, (d), the phase may change so rapidly with frequency that the phase-correction routines which use a short doubled-sided portion of the interferogram may not be long enough to allow the true intensities to be calculated.

For slow-scanning interferometers in the far-infrared (see Section X), the interferogram is usually perfectly symmetrical when the interferometer is in good alignment. As the alignment deteriorates, not only does the amplitude of the signal at zero retardation decrease but the interferogram also becomes asymmetrical. This demonstrates that the zero *phase* difference point no longer occurs at the same *path* difference for all frequencies. For a perfectly aligned slow-scanning interferometer where all frequencies are in-phase at zero retardation, the amplitude of the signal at zero retardation, $I(0)$, should be equal to twice the average value of the signal at high retardation, $I(\infty)$. The ratio, R, where

$$R = \frac{I(0) - I(\infty)}{I(\infty)} \qquad (1.38)$$

gives a measure of how well the interferometer is in alignment. For well-adjusted interferometers, R should be greater than 0.9.

Alignment of the mirrors of an interferometer is almost invariably needed when the beamsplitter is changed. It is occasionally needed at infrequent intervals to peak up performance when the energy at high frequency in the spectrum appears to have decreased. Alignment is generally performed by adjusting the plane of the fixed mirror, so that the angle between the plane of the fixed mirror and the plane of the beamsplitter is the same as that for the moving mirror; at this point the interferogram will be symmetrical about zero retardation. It is not necessary for these angles to be precisely 45°, but it is necessary for them to be the same.

For rapid-scanning interferometers, the interferogram is always slightly chirped and so these interferometers cannot be aligned using the symmetry of the interferogram as a guide. To align a rapid-scanning interferometer, the moving mirror should be scanned repetitively about the zero retardation point and the fixed mirror is adjusted to give the maximum signal. To check that the alignment is good, it is advisable to have a reference single-beam spectrum at hand, taken with the same source, beamsplitter and detector, measured when it is known that the alignment was good. If the energy at high frequency in the current spectrum is lower than the energy at the corresponding frequency in the reference spectrum, then further alignment is necessary.

IX. EFFECT OF A POOR MIRROR DRIVE

The quality of the drive mechanism of the moving mirror ultimately determines whether a certain interferometer can be used to measure a spectrum with a resolution corresponding to the maximum retardation of the interferometer. The retardation only determines the resolution of an interferometer if the mirrors maintain good alignment throughout the entire scan and if the beam passing through the interferometer is sufficiently collimated.

The effect of a drive mechanism that does not allow the plane of the moving mirror of the interferometer to maintain its angle relative to the plane of the beamsplitter is somewhat analogous to the effect of beam-divergence discussed in Section VII. In the case of the poor mirror drive, however, there is an optical path difference generated between the two extreme rays of the beam passing through the interferometer rather than a path difference between the extreme rays and the central ray.

A common type of drive defect occurs when the top of the moving mirror gradually tilts as the retardation increases (Fig. 1.19). Let the mirror tilt at a

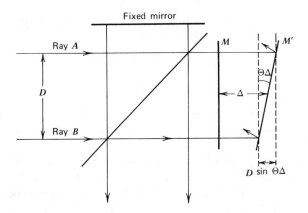

Fig. 1.19. The effect of tilting the moving mirror of the interferometer by $\Theta\Delta$ is to increase the retardation for ray A and decrease it for ray B.

rate of Θ radians per centimeter of retardation. Consider the effect of a collimated beam of radiation of wavelength λ and diameter D cm entering the interferometer. At any retardation, Δ, the increase in retardation, x, for the upper ray (A) over the lower ray (B) is

$$x = 2D \sin (\Theta\Delta). \tag{1.39}$$

If $\Theta\Delta$ is small

$$x = 2D\Theta\Delta,$$

or

$$\Theta = \frac{x}{2D\Delta}. \tag{1.40}$$

Loss of fringe modulation occurs when

$$x = \frac{\lambda}{2} = \frac{1}{2\bar{v}}.$$

Thus, to ensure that no degradation of resolution occurs during the entire scan,

$$\Theta < \frac{1}{4\bar{v}_{max} \cdot D \cdot \Delta_{max}}, \tag{1.41}$$

where $\bar{v}_{max}$ is the highest frequency in the spectrum and Δ_{max} is the maximum retardation. For an ideal interferometer, the resolution $(\Delta\bar{v})$ cm^{-1} is given by approximately $(\Delta_{max})^{-1}$, so that

$$\Theta < \frac{1}{4D} \frac{(\Delta\bar{v})}{\bar{v}_{max}}. \tag{1.42}$$

To illustrate the kind of drive accuracy required for mid-infrared spectroscopy consider the following parameters:

$$\bar{v}_{max} = 5000 \text{ cm}^{-1}$$
$$\Delta\bar{v} = 0.5 \text{ cm}^{-1}$$
$$D = 5 \text{ cm}.$$

When these parameters are inserted into Eq. 1.42, it may be seen that Θ_{max} is 5×10^{-6} radians/cm retardation, or approximately 1 sec of arc/cm retardation.

If the resolution is found to have been degraded because of mirror tilt, the diameter of the beam may be reduced; thus, in the above example, if a resolution of 0.5 cm^{-1} is not attained above 2500 cm^{-1}, the beam may be apertured down to a diameter of 2.5 cm. When the area of the beam is so reduced, however, the amount of energy reaching the detector is reduced in proportion to the area of the beam, that is, by a factor of four, and the signal-to-noise ratio of the spectrum would be reduced by this amount.

X. SLOW-SCAN AND RAPID-SCAN INTERFEROMETERS

In practice, interferometers are used in two different ways depending on the scan speed of the moving mirror.

For *slow-scanning interferometers* the velocity of the moving mirror, V cm/sec, is sufficiently slow that the modulation frequency, $f_{\bar{v}}$, of each of the spectral frequencies, $\bar{v}$, is generally less than 1 Hz, where

$$f_{\bar{v}} = 2V\bar{v} \text{ Hz} \tag{1.10}$$

This type of system is most commonly used for far-infrared spectroscopy, where the highest frequency is generally about 600 cm^{-1}. A typical scan speed is of the order of 4μm/sec, so that the highest frequency in the spectrum has a modulation frequency,

$$f_{600} = 2 \times 4 \times 10^{-4} \times 600$$
$$= 0.5 \text{ Hz}$$

in the interferogram. Far-infrared frequencies fall between about 6 and 600 cm^{-1}, so that if the mirror velocity used in the above example were used for all measurements, the range of modulation frequencies would be between

5×10^{-3} and 5×10^{-1} Hz. In practice, when very far-infrared frequencies are being measured, scan speeds as low as 0.4 μm/sec can be used, so that a lower limit on the modulation frequency is more like 5×10^{-4} Hz.

Frequencies in this subaudio range are rather difficult to amplify without picking up a large amount of $1/f$ noise. As a result the usual technique for measuring far-infrared spectra with a slow-scanning interferometer is to modulate the beam with a mechanical chopper. The frequency of the chopper is designed to be considerably higher than the modulation frequency of the shortest wavelength radiation to be measured. With this system, a lock-in amplifier can be easily used to amplify the signal before digitization.

The signal that is measured in this way consists of both the a.c. and the d.c. portions of the interferogram. Since it is only the a.c. portion that is of interest for the computation of the Fourier transform, the average value of the interferogram must be computed and subtracted from the value of each sample point before the transform is performed.

For accurate sampling of the interferogram, the position of the moving mirror should be monitored by a secondary fringe-reference device in conjunction with the interferometer drive mechanism. These devices generally involve the use of either a Moire fringe system or a laser generating a sinusoidal wave as the mirror travels during the measurement (see Chapter 4 for further details). Using these devices, the value of the interferogram can be measured at equal intervals of retardation rather than equal intervals of time, so that any nonuniformity in the velocity of the moving mirror is compensated for.

An electrical filter circuit is used to effect integration of the signal between sample points. If this circuit were not present, noise of higher frequency than the modulation frequencies of the radiation being measured would be seen in the interferogram. It will be seen in the next chapter that this effect would increase the noise level in the computed spectrum. The time constant of the filter must therefore be set to match the noise bandwidth to the frequency bandwidth of the signal.

Several interferometers of the slow-scanning variety use a *stepping-motor* drive whereby the retardation is increased in a staircase pattern as shown in Fig. 1.20. The interferogram signal is produced by integrating the detector output signal during the time interval that the mirror is held stationary. The time interval during which the mirror is stepped from one sampling position to the next is lost time that does not contribute to the observation and therefore should be kept short relative to the integration periods [12].

The success of the technique depends on the capability of the stepping motor to achieve a uniform increase in retardation for every step. For far-infrared spectroscopy at low resolution, the technique works satisfactorily with a commerical stepping-motor using no additional optics or electronics.

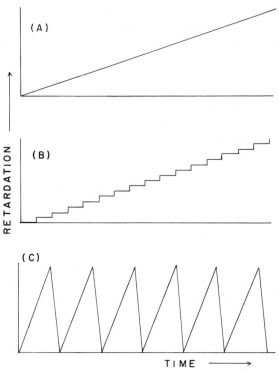

Fig. 1.20. The different drive methods for an interferometer: (a) the continuous slow-scanning drive method, where the retardation increases linearly with time throughout the entire measurement; (b) the step-and-integrate method, where the movable mirror is held at equidistant intervals of retardation for equal periods of time and translated very rapidly between these points; (c) the rapid-scanning method, where the complete interferogram is scanned rapidly but continuously; at the end of each scan, the mirror is quickly retraced for the start of the next scan.

However, the positioning accuracy required for high resolution mid- and near-infrared spectroscopy is such that the mirror position has to be controlled by a servomechanism actuated after every step by a laser reference interferometer.

Rapid-scanning interferometers differ from the slow-scanning variety in that the mirror velocity of the interferometer is sufficiently high that the spectral frequencies are modulated in the audio-frequency range. It will be shown in Chapter 2 that rapid scanning is mandatory for the successful measurement of low or medium resolution mid-infrared absorption spectra, where the interferogram has an extremely high signal-to-noise ratio.

A typical mirror velocity used for the measurement of spectra between 4000 and 400 cm^{-1} is 0.2 cm/sec. The modulation frequencies for the upper and lower limit of this range are, therefore,

$$f_{4000} = 2 \times 0.2 \times 4000$$
$$= 1600 \text{ Hz}$$
$$f_{400} = 160 \text{ Hz}.$$

Since these frequencies are in the audio-range, they can be easily amplified without the necessity for modulating the beam with a chopper. (If such a chopper were used, it would have to modulate the beam at a considerably higher frequency than the modulation frequency of the shortest wavelength being measured, i.e., at least 10 kHz).

The signal from the detector of a rapid-scanning Fourier transform spectrometer must be amplified using a band-pass filter. In this way, noise of higher frequency than the modulation frequency of the shortest wavelength in the spectrum is eliminated, while slow variations in source intensity fall below the lower limit of the band-pass. The measured interferogram represents the a.c. component of the signal, so that there is no need to subtract the mean value of the interferogram from each sample point.

To increase the signal-to-noise ratio (S/N) in the interferogram, signal-averaging techniques must be used. Thus, all corresponding data points must be digitized at exactly the same retardation value on every scan. Several commerical instruments use a combined white-light and laser fringe-referenced interferometer to achieve this coherent addition (see Chapter 4 for details).

Some of the advantages and disadvantages of rapid-scan and slow-scan interferometry are summarized below.

Rapid-scan	Slow-scan
Signal-to-noise ratio in the interferogram *per scan* is sufficiently low that mid-IR spectra can be measured using intense sources by signal-averaging.	S/N is too high for accurate digitization of the noise level in interferograms of intense sources, so that mid-IR absorption spectra cannot be measured efficiently.
Rather high scan speeds are necessary to generate modulation frequencies > 10 Hz for very long wavelengths in the far-infrared.	Well suited to far infrared spectroscopy.
Observation efficiency is reduced by the time taken to retrace the moving mirror during signal averaging.	Observation efficiency for step-and-integrate systems is reduced by the time taken to increment the retardation between sample points.

Rapid-scan	Slow-scan
Unaffected by slow variations in source intensity.	Slow variations in the source intensity are measured, resulting in degradation of the spectra especially at low frequency.
No chopper needed; therefore all the radiation hits the detector all of the time.	Chopper is generally needed (except when phase modulation techniques are used); therefore on the average half the signal is lost.
Necessitates the use of rapid-response detectors.	Can use detectors with a long response time.
Increasing the S/N by signal averaging in real-time necessitates the use of a hardwired signal-averager or minicomputer. However, if a minicomputer is used it can also be used to perform the transform immediately after the measurement.	Recording the interferogram can be carried out using less expensive equipment. But, if a computer is not part of the spectrometer, the transform must be performed on a remote computer (with the ensuing costs).
Only one scan-speed need be used, and therefore the relationship between S/N, measurement time and resolution (the "trading rules") is easily calculated.	The "trading rules" are more difficult to calculate. When the scan speed is changed other instrumental parameters (e.g. the filter time constant) must also be changed.
"Real-time" computations are very difficult to perform at the current state-of-the-art.	"Real-time" computations can be performed (see Chapter 3).
Interferograms are usually asymmetric, and phase correction must be performed.	Interferograms are usually symmetric, so that phase correction is often unnecessary. (see Section VI)
Not applicable to ultrahigh resolution ($<.01$ cm^{-1}) measurements.	Have been successfully applied to high resolution measurements (see Chapter 4).

In summary the choice of interferometer should be made on the basis of the measurements to be made. For medium resolution mid-infrared absorption spectroscopy, the use of a rapid scan system is mandatory when the energy reaching the detector is high. On the other hand, if very high resolution measurements are desired, a slow-scanning interferometer using the step-and-integrate technique is desirable. For far-infrared spectroscopy, either type of spectrometer can be used.

XI. PHASE MODULATION

One of the advantages of rapid-scanning interferometry over slow-scanning interferometry is that no chopper is used to modulate the radiation, which results in a gain of a factor of two for corresponding measurements. Another disadvantage of slow-scanning interferometers is that slow variations in the intensity of the source can result in variations of the baseline of the interferogram which can be of the same frequency as the modulations from the longest wavelength being measured. This effect can give very poor photometric performance at these low frequencies; it is often noticed in far-infrared spectroscopy where the mercury arc source generally used for these measurements is rather unstable so that its intensity can drift quite badly. The effect is also prevalent in astronomical spectroscopy where the intensity of distant sources may change slowly due to passing clouds, poor following, fluctuations of atmospheric transparency or fluctuations of refractive index (scintillation). One method of getting around the problem of slowly varying sources while at the same time picking up approximately a factor of two in signal-to-noise ratio by eliminating the chopper is to use a technique known as *phase modulation* (PM).

This technique has been independently developed by Chamberlain's group [13, 14] for far-infrared spectroscopy and by Connes et al. [15, 16] for near-infrared measurements. The method (called internal modulation by Connes) requires the use of a slow-scanning interferometer usually employing the step-and-integrate method of traversing the moving mirror. At each step the beam is modulated not by a chopper (called amplitude modulation (AM) by Chamberlain) but by periodically varying the retardation by a small amount.

The technique used by Connes et al. [16] for this purpose is to support a pair of transparent plates in each arm of the interferometer; one pair is fixed and the other pair is mounted on a driven tuning fork. Oscillation of this tuning fork gives a small periodic variation of path difference within the interferometer and thus modulates the output. Chamberlain could not use this technique because of the difficulty of obtaining media for the modulating element that are transparent and nondispersive over the far-infrared spectral range. They instead modulate the path difference by a small periodic displacement of one of the interferometer mirrors. The amplitude of the oscillation determines the efficiency of the technique [13], and it depends on the wavelength of the radiation being measured, far-infrared measurements requiring substantially greater displacements than mid- or near-infrared spectroscopy.

To illustrate the difference between amplitude and phase modulation, let us consider an idealized spectrum of the type shown in Fig. 1.21. The interferogram when AM is used for the measurement is a sinc function of the type shown in Fig. 1.22a. However, if phase modulation is used, the measured signal represents the change of this AM signal over a small increment of retardation. If the amplitude of the jitter were infinitesimally small, the

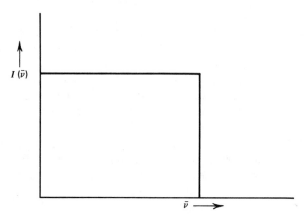

Fig. 1.21. Synthetic "white" band-limited spectrum. (Reproduced from [13] by permission of the author and Pergamon Press; copyright ©️ 1971.)

signal would be the first derivative of the AM interferogram. In practice, even with finite jitter amplitudes, the PM interferogram is to a very good approximation the first derivative of the AM interferogram. Thus the PM interferogram of the spectrum shown in Fig. 1.21 would have the form shown in Fig. 1.22b.

Consider the case of a monochromatic source. The AM interferogram is given by Eq. 1.2, that is,

$$I'(\delta) = 0.5\ I(\bar{\nu})\{1 + \cos 2\pi\bar{\nu}\delta\}. \tag{1.2}$$

The derivative of this gives the PM interferogram, $I''(\delta)$:

$$I''(\delta) = \frac{d}{dt}\left[I'(\delta)\right] = 0.5\ I(\bar{\nu})\sin 2\pi\bar{\nu}\delta. \tag{1.43}$$

Thus the spectrum can be computed from an interferogram measured using PM by the sine transform.

It is worthwhile to compare the results shown in Fig. 1.22 to Fig. 1.10. In the latter case, a sampling error of $\lambda/4$ in an AM interferogram has caused

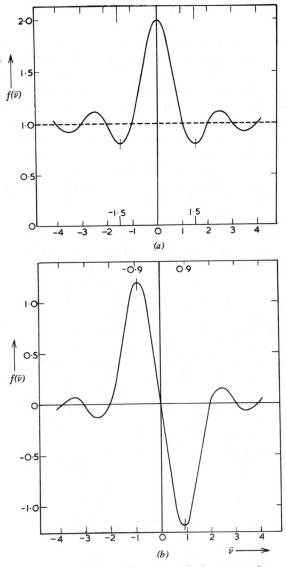

Fig. 1.22. Interferograms from the synthetic spectrum shown in Fig. 1.21; (a) measured with amplitude modulation, and (b) measured with phase modulation. (Reproduced from [13] by permission of the author and Pergamon Press; copyright ⓒ 1971.)

the interferogram to appear to be a sine wave rather than a cosine wave, resulting in an instrumental line shape identical to the shape of the interferogram in Fig. 1.22b when the cosine transform is performed.

In practice interferograms measured using AM and PM have a shape quite similar to those shown in Fig. 1.22 and typical results for the far-infrared region are shown in Fig. 1.23. To perform these measurements, Chamberlain [14] replaced the moving mirror by a "jittered" mirror which is a thin plane mirror mounted on the plunger of a small loudspeaker coil. The sinusoidal signal needed to drive the mirror is obtained by a sine-wave generator. The zero retardation position is first found using a chopper, and then the chopper is stopped and the jitter is started using a frequency chosen to be the same as that of the chopper to allow valid comparisons to be made between the AM and PM measurements. Since the magnitude of the PM interferogram is zero at zero retardation, the position of the movable mirror must be adjusted to the position where the PM interferogram has its maximum. The amplitude of the jitter is then increased until the greatest signal at the detector is developed. The amplitude should be approximately one-quarter of the central wavelength in the spectral range being measured (which is about one-half of the shortest wavelength in the range). For good performance, it is essential that the equilibrium position of the vibrator remains unchanged during the measurement, that the amplitude of the jitter remains constant and that the tilt of the vibrating mirror does not alter.

Connes' group has also used an analogous technique for near-infrared measurements using a piezo-electric element to provide the high frequency, low amplitude jitter required for this spectral range. This technique is not feasible for far-infrared spectroscopy since a sufficiently large amplitude cannot be achieved with a piezo-electric element for this region.

Figure 1.23 shows AM and PM interferograms of the same source measured by Chamberlain and Gebbie [14] and Fig. 1.24 shows the transform of these interferograms shown on the same (arbitrary) scale when the interferograms were recorded with the same gain. It can be seen that the amplitude of the spectrum measured using phase modulation is about twice as great as that of the spectrum measured using amplitude modulation, which demonstrates the factor of two gain obtained on removing the chopper. Although it is rather difficult to see from these spectra, the noise in each spectrum is essentially the same.

The efficiency of PM measurements relative to AM measurements falls off at the extremes of the spectral range being measured. This effect is predominantly due to the fact that the modulation of a cosine function (in this case the interferogram) by another cosine function (the modulation of the mirror) generates Bessel coefficients. Since only the lowest order coefficient is detected, energy is lost in the higher order coefficients.

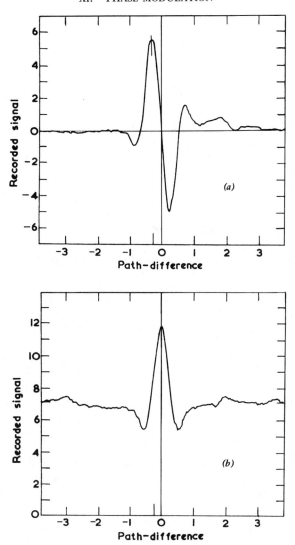

Fig. 1.23. Actual interferograms of the same source measured with (a) phase modulation and (b) amplitude modulation. (Reproduced from [14] by permission of the author and Pergamon Press; copyright © 1971.)

It will be seen in Chapter 4, Section II (Beamsplitters) that the profile of the spectrum in Fig. 1.24 is typical of spectra measured with stretched-film beamsplitters—the type most commonly used for far-infrared spectroscopy. These beamsplitters can be used to measure spectra over a frequency range

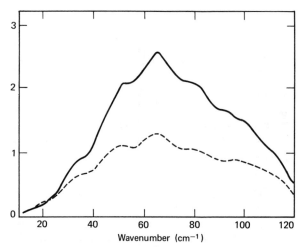

Fig. 1.24. The transform of the previous interferograms shown
on the same scale. The broken line is from the amplitude-
modulated interferogram and the solid line is from the phase-
modulated interferogram. (Reproduced from [14] by permission
of the author and Pergamon Press; copyright © 1971.)

0 to $\bar{v}_M$ cm^{-1}. However the energy at the low and high frequency extremes
of this range is always low relative to the central region, thus limiting the
useful range to about $0.15\bar{v}_M$ to $0.85\bar{v}_M$. Chamberlain [10] has examined
the ratio, R, of the theoretical signal-to-noise ratio in spectra measured
using PM and AM, when the only source of noise in the interferogram is
detector noise; the results are shown in Fig. 1.25. It is seen that R is greater
than unity for $0.13\bar{v}_M < \bar{v} < 0.89\bar{v}_M$. Thus, for the total useful range of the
far infrared, using stretched-film beamsplitters, phase modulation does
indeed give a decided advantage over amplitude modulation techniques.
This result has been demonstrated experimentally by Chamberlain and
Gebbie [11].

An experimental comparison between PM techniques with a slow-scanning
interferometer and AM techniques with a rapid-scanning interferometer has
not yet been made, but the two methods should give rather similar results
in terms of signal-to-noise ratio since neither technique uses a chopper.

Since the mirror jitter has a frequency greater than 1 Hz, the effect of low-
frequency source variations is essentially eliminated using PM techniques
so that the low frequency performance should be substantially superior to
that of the corresponding AM measurement using a slow-scanning interferometer. One other advantage is also found which is related to the same

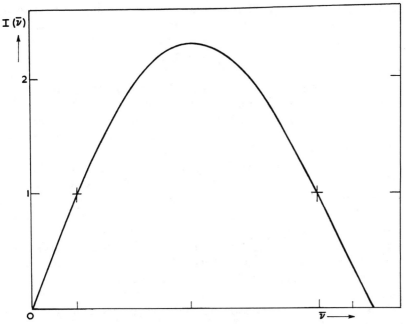

Fig. 1.25.   Ratio of the signal-to-noise ratios obtained from a given source using PM and AM. (Reproduced from [13] by permission of the author and Pergamon Press; copyright ⓒ 1971.)

cause. Since a PM interferogram has no d.c. component, it has a more useful dynamic range for digitization so that it is easier to extract all the information required for the spectrum from a given interferogram. Both of these advantages of PM over AM interferometry using a slow-scanning interferometer have also been shown to hold true for a rapid-scan over a slow-scan interferometer, so that the advantage of rapid-scanning interferometry over PM methods (or vice versa) is as yet still undecided.

XII.   REFRACTIVE INDEX MEASUREMENTS

For most measurements that have been made using a Michelson interferometer, the sample is held either between the source and the interferometer or between the interferometer and the detector. Under "ideal" conditions the interferogram measured in this way is perfectly symmetrical on either side of zero retardation, that is, $\theta_{\bar{\nu}} = 0$; most interferograms measured using slow-scanning interferometers for far-infrared spectroscopy are indeed very symmetrical. However, if a sample of thickness d is placed in one *arm* of the interferometer, the interferogram no longer retains this symmetry unless the refractive index of the sample, $n(\bar{v})$, is exactly the same across the complete frequency range being measured since the optical path-length in that arm is increased by an amount equal to $2d\{n(\bar{v}) - 1\}$.

If $n(\bar{v})$ is constant throughout the spectral range being studied, the only effect on the interferogram is to displace the position of zero retardation. On the other hand if $n(\bar{v})$ is frequency dependent, the zero phase difference point will occur at different mirror positions for different frequencies. This effect is analogous to have a large frequency dependent phase angle, $\theta_{\bar{v}}$, as discussed in Section VI.

Bell [17] has published two interferograms which nicely illustrate this effect and they are reproduced in Fig. 1.26. The refractive index of polyethylene is essentially constant throughout the frequency range under study, and this is reflected in the fact that the interferogram is merely shifted with respect to the interferogram measured with no sample in the interferometer but it is not distorted. On the other hand when a sheet of Mylar is placed at the same position, the interferogram becomes asymmetrical, demonstrating that the refractive index of Mylar in the far infrared varies with frequency.

When monochromatic radiation of frequency $\bar{v}_0$ is passed into a Michelson interferometer, the interferogram is a cosine wave. When a sample is inserted into one arm of the interferometer, the amplitude of the cosine wave is reduced due to absorption by the sample, and the zero retardation fringe is displaced by an amount $2d\{n(\bar{v}_0) - 1\}$, see Fig. 1.27. The assignment of any one particular fringe to the zero retardation position is impossible in this case since each fringe has the same amplitude. Thus, unless further information is available, the assignment of the refractive index of a sample measured under these conditions cannot be done. Two techniques [18] have been used to get around this problem.

The first method involves the use of "quasi-monochromatic" radiation, that is radiation of narrow but finite band width. Under these circumstances the zero retardation fringe can be recognized, see Fig. 1.28. This distribution can be achieved from a broad-band source using a narrow band-pass filter. In view of the narrow width of the pass-band, the value of $n(\bar{v})$ of the material being studied can be assumed to be constant across the frequency range transmitted through the filter, so that $n(\bar{v}_0)$ can be obtained accurately from the displacement of the zero retardation fringe. This method has the disadvantage of requiring as many filters as frequencies necessary for a plot of the refractive index spectrum.

A method for determining $n(\bar{v}_0)$ at the frequency of a strictly monochromatic light source has been developed by Chamberlain and Gebbie [18] in which the sample is rotated in the arm of the interferometer. By setting

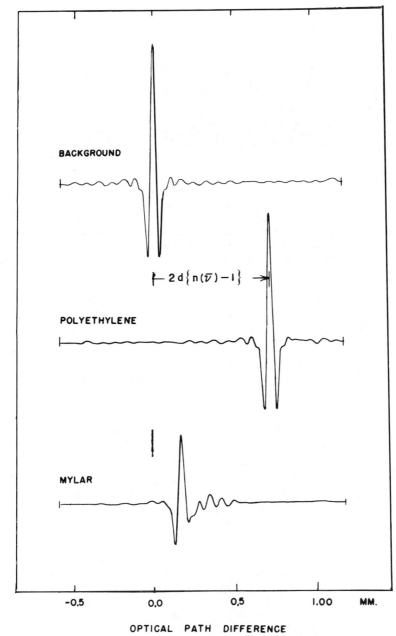

Fig. 1.26. The central portions of a background interferogram, an interferogram measured with a 1.51 mm thick polyethylene sample in one arm of the interferometer and an interferogram measured with a 0.25 mm thick Mylar film in one arm of the interferometer [17].

53

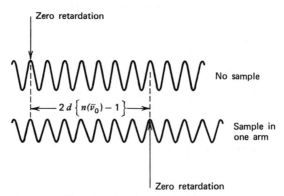

Fig. 1.27. The effect on the interferogram of placing a sample of thickness, d, and refractive index, $n(\bar{\nu})$, in one arm of the interferometer for monochromatic radiation of frequency $\bar{\nu}_0$. Note that the zero retardation fringe cannot be distinguished from any of the other maxima.

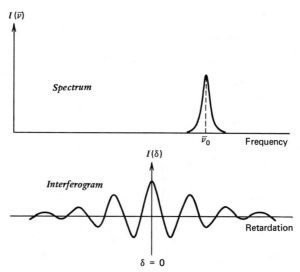

Fig. 1.28. (a) A spectrum of "quasi-monochromatic" radiation; (b) the interferogram from this spectrum; the zero retardation fringe can now be distinguished and the bandwidth of the spectrum is small enough that the refractive index can be considered to be constant across this range.

54

the movable mirror of the interferometer stationary at such a position that a bright fringe is observed at the detector with the sample normal to the beam and then recording the angles to which the specimen has to be rotated for other bright fringes to be observed, very accurate values for $n(\bar{v}_0)$ were able to be calculated. The single obvious disadvantage of this method is that a source of monochromatic radiation is needed for every frequency at which the refractive index is to be determined. For this reason, Chantry and Chamberlain [19] have only given results for a single wavelength in the far infrared, that of the 337 μm HCN laser.

It is obviously preferable to be able to compute the complete refractive index spectrum from an interferogram measured using a broad-band source with the sample held in one arm of the interferometer (the type of interferogram shown in Fig. 1.26). However, computing the refractive index spectrum from an asymmetric interferogram is by no means as simple as computing the power spectrum from a symmetrical interferogram. The theory behind *Fourier refractometry* can be found in several papers. A rigorous description of this theory is beyond the scope of this book but a limited explanation will be given at this point.

The modulated portion of the interferogram measured at a retardation δ is given in the absence of the specimen by Eq. 1.23, that is,

$$I(\delta) = \int_0^{+\infty} B(\bar{v}) \cos(2\pi\bar{v}\delta - \theta_{\bar{v}}) \cdot d\bar{v}.$$

When a sample is present in one arm of the interferometer, it is most desirable for the sake of computational convenience to have the origin of the calculations as the point where the maximum amplitude of the signal is measured [19]. If this point occurs at a distance $\bar{\delta}$ from the corresponding point when no sample was present in the beam, we can define a new variable, γ, where

$$\gamma = \delta - \bar{\delta}, \tag{1.44}$$

so that the interferogram is given by

$$I'(\gamma) = \int_0^{+\infty} B(v) \cos\{2\pi\bar{v}\gamma + \Psi(\gamma)\} \cdot d\bar{v}, \tag{1.45}$$

where

$$\Psi(\gamma) = \theta_{\bar{v}} - 2\pi\bar{v}\delta. \tag{1.46}$$

Provided that the data is collected "double-sided" (see Section VI) we may take the cosine transform, $T_{\cos}(\bar{v})$, and the sine transform, $T_{\sin}(\bar{v})$, so that the power spectrum, $|\hat{S}(\bar{v})|$, given by Eq. 1.25 as

$$|\hat{S}(\bar{v})| = \{[T_{\cos}(\bar{v})]^2 + [T_{\sin}(\bar{v})]^2\}^{\frac{1}{2}} \tag{1.47}$$

while the phase $\Psi(\bar{v})$ is derived from

$$\text{ph} . \hat{S}(\bar{v}) = -\psi(\bar{v}), \tag{1.48}$$

where $\psi(\bar{v})$ is the principal value of $\Psi(\bar{v})$ by noting that

$$\Psi(\bar{v}) = \psi(\bar{v}) + 2M\pi \qquad (M = 0, 1, 2 \text{ etc.}). \tag{1.49}$$

If another interferogram is measured with no specimen in the interferometer, $\bar{\delta} \to 0$ and $\gamma \to \delta$, so that the transformation is carried out about $\delta = 0$ as origin. In this case we obtain:

$$|\hat{S}_o(\bar{v})| = \{[t_{cos}(\bar{v})]^2 + [t_{sin}(\bar{v})]^2\}^{\frac{1}{2}} \tag{1.50}$$

while the phase angle, this time denoted by $\Phi(\bar{v})$ is derived from

$$\text{ph} \cdot S_0(\bar{v}) = -\phi_0(\bar{v}) \tag{1.51}$$

where $\phi_0(\bar{v})$ is the principal value of the phase $\Phi_0(\bar{v})$, that is,

$$\Phi_0(\bar{v}) = \phi_0(\bar{v}) + 2m_0\pi. \tag{1.52}$$

From these equations, it may be shown that the refractive index at any frequency is given by

$$n(\bar{v}) = 1 + \frac{\bar{\delta}}{2d} + \frac{1}{4\pi\bar{v}\delta} \{\text{ph} \cdot \hat{S}(\bar{v}) - \text{ph} \cdot S_0(\bar{v}) + 2(M - m_0)\pi\} \tag{1.53}$$

If $n(\bar{v})$ has a constant value, the last term becomes equal to zero, so that

$$n(\bar{v}) = 1 + \frac{\bar{\delta}}{2d} \cdot \tag{1.54}$$

This method of refractometry is sometimes called "amplitude spectroscopy" since the magnitude and phase changes in the electromagnetic waves incident upon a sample are measured, as opposed to "power spectroscopy" where the power changes in the incident wave are measured. More often it is referred to as "asymmetrical interferometry" since the symmetry of the interferogram measured using this technique is almost invariably reduced.

The problems of asymmetric interferometry are increased by the multiple reflections that can occur in the plane-parallel samples necessary for refractometry. Fig. 1.29 shows the various rays that can reach the detector from such a sample. The higher the refractive index of the material, the smaller is the proportion of the primary rays (shown with a heavy line) to the secondary rays. Each of the secondary rays will interfere with the beam from the other arm of the interferometer to produce its own interferogram, or *signature*. To eliminate one of these secondary rays, Chamberlain et al. [20, 21] developed a method by which the refractive index of liquids could be measured without using a cell, see Fig. 1.30. By holding the fixed mirror in a horizontal plane, a film of liquid of any thickness can be retained on the surface of the mirror.

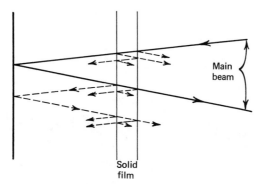

Fig. 1.29. The beams that can reach the detector when a sample is inserted in one arm of the interferometer caused by reflections from various surfaces of the sample. Only the beam indicated by the solid line is of interest.

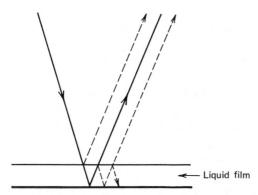

Fig. 1.30. The number of unwanted beams is reduced for liquid samples if the liquid is held on the fixed mirror, provided that this mirror is held in a horizontal plane.

Even with this arrangement, secondary signatures are still found due to waves reflected from the upper surface of the liquid and due to multiple internal reflections. Figure 1.31. shows an asymmetrical interferogram measured by Chamberlain et al. [20] for a film of tetrabromoethane of 0.68 mm thickness. To simplify the computations, Chamberlain removed the signature due to the first surface reflection merely by replacing the values of the interferogram in the region of this signature by zero values. The signatures caused by internal reflection were sufficiently weak that they did not affect the accuracy of the measurement significantly.

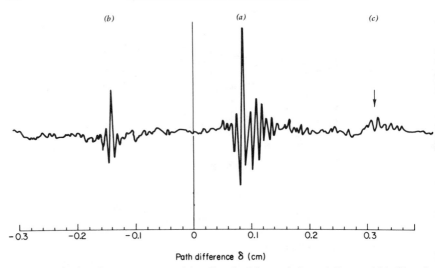

Fig. 1.31. An interferogram measured by Chamberlain et al. for a 0.68 mm thick film of liquid sym-tetrabromoethane held as in Fig. 1.30. The signature (b) is due to the partial wave reflected from the top surface of the liquid while (c) is due to the partial wave that has been internally reflected once from the upper surface of the liquid. (Reproduced from [20] by permission of the author and Pergamon Press; copyright © 1967.)

If the transmittance of the material is high, the thickness of the film may be chosen to be large enough so that the signatures do not occur in the region of the interferogram containing the primary interferogram over which the computations are going to be performed. Naturally this technique would preclude the use of high resolution measurements where the length of the interferogram being studied is great.

Sanderson [22, 23] has measured the refractive index of gases in the region of their pure rotational spectrum by placing identical gas cells in each arm of the interferometer. This technique cannot be used for the measurement of the refractive index of liquids since the proportion of radiation reflected from each surface of the windows in contact with the liquid would be changed, since the refractive index of the liquid will be close to that of the window material.

REFERENCES

1. A. A. Michelson, *Phil. Mag.*, Ser. 5, **31**, 256 (1891).
2. A. A. Michelson, *Phil. Mag.*, Ser. 5, **34**, 280 (1892).
3. A. A. Michelson, *Light Waves and Their Uses*, University of Chicago Press, Chicago 1902 (reprinted, Phoenix Science Series, University of Chicago Press, 1962).

4. R. Bracewell, *The Fourier Transform and its Applications*, McGraw-Hill, New York, 1965.

5. D. C. Champeney, *Fourier Transforms and their Physical Applications*, in G. K. T. Conn and K. R. Coleman, Eds., *Techniques of Physics*, *Vol. 1*, Academic Press, London, 1973.

6. E. G. Codding and G. Horlick, *Appl. Spectrosc.*, **27**, 85 (1973).

7. J. Connes, *Rev. Opt.* **40**, 45, 116, 171, 233 (1961). English translation as Document AD 409869, Clearinghouse for Federal Scientific and Technical Information, Cameron Station, Va.

8. L. Mertz, *J. Phys.*, **28**, C2:11 (1967).

9. L. Mertz, *Infrared Phys.*, **7**, 17 (1967).

10. M. L. Forman, W. H. Steel, and G. A. Vanasse, *J. Opt. Soc. Am.*, **56**, 59 (1966).

11. E. V. Loewenstein, Aspen Int. Conf. on Fourier Spectrosc., 1970, G. A. Vanasse, A. T. Stair, and D. J. Baker, Eds., AFCRL–71–0019, p. 3.

12. H. Sakai, Aspen Int. Conf. on Fourier Spectrosc., 1970, G. A. Vanasse, A. T. Stair, and D. J. Baker, Eds., AFCRL–71–0019, p. 19.

13. J. Chamberlain, *Infrared Phys.*, **11**, 25 (1971).

14. J. Chamberlain and H. A. Gebbie, *Infrared Phys.*, **11**, 57 (1971).

15. J. Connes, P. Connes, and J-P. Maillard, *J. Phys.*, **28**, C2:120 (1967).

16. J. Connes et al., *Nouvelle Rev. Opt.*, Appl. No. 1 (1970).

17. E. E. Bell, Aspen Int. Conf. on Fourier Spectrosc., 1970, G. A. Vanasse, A. T. Stair, and D. J. Baker, Eds., AFCRL–71–0019, p. 71.

18. J. Chamberlain and H. A. Gebbie, *Nature*, **206**, 602 (1965).

19. G. W. Chantry and J. Chamberlain, Chapter 20 in A. D. Jenkins, Ed., *Polymer Science*, North-Holland, Amsterdam, 1972.

20. J. Chamberlain, A. E. Costley, and H. A. Gebbie, *Spectrochim. Acta*, **23A**, 2255 (1967).

21. J. Chamberlain, H. A. Gebbie, G. W. F. Pardoe, and M. Davies, *Chem. Phys. Letters*, **1**, 523 (1968).

22. R. B. Sanderson, *Appl. Opt.*, **6**, 1527 (1967).

23. W. H. Robinette and R. B. Sanderson, *Appl. Opt.*, **8**, 711 (1969).

SAMPLING THE INTERFEROGRAM

I. SAMPLING FREQUENCY

It was seen from Eqs. 1.11 and 1.12 that in order to compute the complete spectrum from 0 to ∞ cm^{-1}, we would have to digitize the interferogram at infinitesimally small increments of retardation. This is, of course, impossible, since in order to calculate the Fourier transform with a reasonably short computing time a limited number of data points must be sampled from the analog interferogram. Just how often the interferogram should be digitized is a problem that has been solved by information theoreticians.

Any waveform that is a sinusoidal function of time or distance can be digitized unambiguously using a sampling frequency equal to twice the bandwidth of the system [1]. The signal (or interferogram in our case) may then be completely reconstructed *without any loss of information or signal-to-noise ratio*. Consider the case of a spectrum in which the highest frequency reaching the detector is $\bar{v}_{max}$ cm^{-1}. The frequency of the cosine wave in the interferogram corresponding to $\bar{v}_{max}$ is $2V\bar{v}_{max}$ Hz (from Eq. 1.10) and therefore the interferogram must be digitized at a frequency of $4V\bar{v}_{max}$ Hz or once every $(4V\bar{v}_{max})^{-1}$ seconds. This is equivalent to digitizing the signal at retardation intervals of $(2\bar{v}_{max})^{-1}$ cm. It is better to design a system where the signal is sampled at equal intervals of retardation rather than equal intervals of time, since if the mirror velocity varies slightly during the scan, the signal is still sampled at the correct intervals of retardation. Using this type of sampling, only gross deviations in mirror velocity cause any deleterious effects in the computed spectrum.

When a sampling frequency of *less* than twice the modulation frequency of any spectral feature is used, the feature at this frequency is computed to occur at another, lower, frequency in addition to its true frequency. Figure 2.1a shows a sinusoidal wave digitized using an allowed sampling interval. It can be seen that this interval is slightly less than one half wavelength. On the other hand, if only every second point were sampled, not only could the original sine wave be drawn through these points, but also a lower frequency wave, Fig. 2.1b. Thus, unless the wave of the highest frequency present in the interferogram is sampled at least twice per wavelength, the information from the high frequency portion of this spectrum will be computed to occur at another lower frequency as well as its true frequency. This phenomenon is known as "folding" or "aliasing".

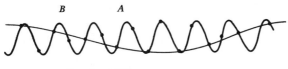

B   A

—•— Sampling points

Fig. 2.1. (a) A correctly sampled sinusoidal wave; the sampling interval is less than one half of the wavelength; (b) if only every second point from (a) is used to digitize the sinusoid, a lower frequency sine wave can also be drawn through these points.

It can also be seen that a higher frequency wave (in fact, many higher frequency waves) can be drawn through the data points in Fig. 2.1a, see Fig. 2.2. No data system can actually distinguish whether the true wavelength when sampled with the frequency shown in Fig. 2.1a is that shown in Fig. 2.1a or Fig. 2.2. It is therefore most important to either limit the range of frequencies hitting the detector by means of optical filters or to limit the bandwidth for which the detector or amplifier has any response.

—•— Sampling points

Fig. 2.2. Higher frequency waves can also be drawn through the wave shown in Fig. 2.1a (solid line in this figure). If the frequency of the solid line is f and the sampling frequency is $2F$, waves of frequency $(2NF \pm f)$ may also be drawn through these points (N is an integer).

If the sampling frequency is $2F$, all the frequencies below F will be transferred through the sampling process unambiguously. On the other hand a feature whose frequency is higher than F will appear at a frequency below F. Figure 2.3 shows a spectrum in schematic form where $2F$ is the sampling frequency [2]. Two features L and M are shown at frequencies greater than F. The features L' and M' show where these features are also computed to occur because of folding. It can be seen that there is no way to distinguish which of the features is real and which arises because of folding. Thus, if the sampling frequency is $2F$, any frequency region of bandwidth F can be examined.

This statement has great importance for high resolution spectroscopy in the near infrared. By limiting the bandwidth of the spectrum with an optical

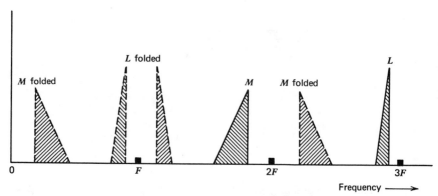

Fig. 2.3. Regions of folding: Two spectral features L and M having frequencies greater than F are also computed to occur at frequencies less than F (Reproduced from [2] by permission of Marcel Dekker, Inc.; copyright © 1972.)

filter, a lower sampling frequency can be used than if the complete spectrum from zero to $\bar{v}_{max}$ were measured. This cuts down the number of data points that have to be collected for a given retardation, and therefore cuts down the computer time required for the Fourier transform. When a small computer is being used for data collection and computation, restriction of the spectral range and the use of a longer sampling interval than $(2\bar{v}_{max})^{-1}$ cm often allows measurements to be taken at a resolution that would otherwise have been beyond the scope of the data system.

In summary, when all the spectrum from zero to $\bar{v}_{max}$ cm^{-1} is required at a resolution of $\Delta\bar{v}$ cm^{-1}, the number of points to be sampled, N_s, is given by

$$N_s = \frac{2\bar{v}_{max}}{\Delta\bar{v}}. \tag{2.1}$$

If the spectral range is restricted to fall between a minimum frequency, $\bar{v}_{min}$, and a maximum frequency, $\bar{v}_{max}$, the number of points required is

$$N_s = \frac{2(\bar{v}_{max} - \bar{v}_{min})}{\Delta v}. \tag{2.2}$$

One very important fact must be remembered when sampling considerations are being discussed. A data system cannot distinguish between signal and noise frequencies, so that any noise whose frequency falls outside the range $2V\bar{v}_{min}$ to $2V\bar{v}_{max}$ Hz will be treated in the same fashion as any real signal outside the range and will be folded back into the spectrum. Thus the noise between $\bar{v}_{min}$ and $\bar{v}_{max}$ will increase. It is therefore most important to restrict the bandwidth of the amplifier so that only the frequency range of interest is studied, and all the other electrical frequencies (whether they correspond to signal or noise) be filtered out. In practice it can be quite difficult to design a filter with a very sharp cutoff, and so generally a slightly higher sampling frequency than required by $\bar{v}_{max}$ and $\bar{v}_{min}$ is used. In this way any noise that is folded back into the spectrum is folded into a region where there is known to be no real spectral information.

An example of a real system where these factors have been brought into consideration could at one time be found in the Model FTS–14 Fourier transform spectrometer made by Digilab Inc [3]. In this spectrometer both optical and electronic filters were used to restrict the bandwidth of the detector output. The sampling interval for this interferometer is determined by multiples of the 632.8 nm line from a helium-neon laser (see Chapter 4, Section I.C). To keep the number of data points for a given resolution at a minimum, the largest possible sampling interval was used; an interval of 2×632.8 nm gives the shortest wavelength allowed in the spectrum as 4×632.8 nm, which corresponds to a frequency of 3951 cm^{-1}. In order that no high *spectral* frequencies are folded back into the spectrum, an optical filter having the profile shown in Fig. 2.4a was placed in the beam. The

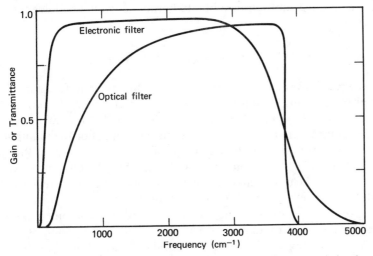

Fig. 2.4. The filters used in the Digilab FTS–14 spectrometer to maximize the sampling interval; (a) the optical filter, which can be made with a sharp cut-off, and (b) the electronic filter for which the cut-off is much less sharp.

mirror velocity of this interferometer is such that 3951 cm^{-1} radiation is modulated at a frequency of 1.25 k Hz, and an electronic filter must be used to cut off noise frequencies greater than this value. The electronic filter used in the FTS–14 has the approximate response shown in Fig. 2.4b, so that even though the optical filter allows no *signal* to be folded back into the spectrum, a small amount of *noise* is folded into the high frequency end of the spectrum when a sampling interval of 2 × 632.8 nm is used. Both to prevent the noise from being folded in this fashion and also to increase the energy at the low frequency end of the spectrum, the optical filter has been eliminated and a sampling interval of 632.8 nm is now used routinely for most interferometers where a helium-neon laser is used to determine the sampling interval.

Another important aspect to be considered is any possible nonlinearity of the detection system. It was stated that any waveform must be digitized at twice the bandwidth of the system so that the signal may be completely recovered. If harmonics of the signal are generated in the detection system, such that the harmonics are outside the bandwidth allowed by the sampling frequency, they will naturally be folded into the real spectrum thereby causing either spurious lines to appear (in the case of a line spectrum) or causing photometric inaccuracy (when the source is continuous). The most drastic case is that of a square wave, which can be expressed as a series:

$$\sin x + \tfrac{1}{3}\sin 3x + \tfrac{1}{5}\sin 5x + \cdots.$$

In this case the complete range of odd harmonics would be folded into the spectrum.

II. DYNAMIC RANGE

In the previous section we discussed how often the interferogram should be digitized; in this section we discuss to what accuracy the amplitude of the signal should be sampled. In view of the fact that the intensity of the interferogram at zero retardation represents the summation of the amplitudes of all the waves in the interferogram, the signal-to-noise ratio at this point can be extremely high when an incandescent broadband source is being measured. On the other hand, the signal-to-noise ratio in the regions of the interferogram well displaced from zero retardation can be rather low.

To give an example of the range of signals found in interferometry, consider the case of the interferogram of an incandescent black-body source generated by a rapid-scanning interferometer and detected by a mid-infrared bolometer. The ratio of the intensity of the signal at zero retardation to the root-mean-square noise level (often called the *dynamic range*) can be as high as 10^4:1. State-of-the-art analog-to-digital converters (ADC's) have a resolution of approximately 15 bits, which means that the signal can be divided up into a maximum of 2^{15} (32768) levels. Thus, if the dynamic range of the interferogram is 10,000, only the two least significant bits of the ADC would be used to digitize the noise level. If the dynamic range of the interferogram were, say, an order of magnitude higher, the noise level would fall below the least significant bit of the ADC and real information would be lost from the interferogram.

This example illustrates an important point for the sampling of interferograms, in that at least two or three bits of the ADC should be sampling detector noise. If no detector noise is seen, the inaccurate sampling of the interferogram can lead to the generation of another type of noise in the spectrum called quantization or *digitization noise*. The source of digitization noise is illustrated in Fig. 2.5, where it can be seen that there is a difference

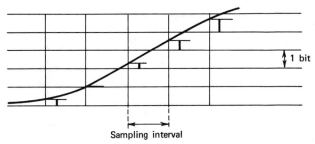

Fig. 2.5. The effect of poor digitization accuracy: The difference between the actual and sampled signals is shown for each data point. This error can be considered to be equivalent to a peak-to-peak detector noise equal to 1-bit.

between the sampled values of the waveform and the true values. This is equivalent to noise in the signal and will obviously transform into spectral noise.

For a spectrum measured between $\bar{v}_{max}$ and $\bar{v}_{min}$ at a resolution of $\Delta\bar{v}$, there are N resolution elements, where

$$N = \frac{(\bar{v}_{max} - \bar{v}_{min})}{\Delta\bar{v}} \qquad (2.3)$$

The dynamic range of the interferogram is given to a good approximation by the dynamic range of the spectrum multiplied by $\sqrt{N}$. For an average spectral dynamic range of, say, 300:1 at a resolution of 2 cm^{-1} for a bandwidth of 4000 cm^{-1}, the dynamic range of the interferogram is $(2000)^{\frac{1}{2}} \times 300$ or 13,500:1. Thus at least a 14-bit ADC would be needed to sample the signal.

The effect of grossly insufficiently resolution of an ADC has been demonstrated by Horlick and Malmstadt [4] using a low resolution single-beam absorption spectrum of a polystyrene film. Their data are reproduced in Fig. 2.6. It is seen that as the resolution of the ADC is decreased, both the frequency resolution and intensity information are degraded in the spectrum.

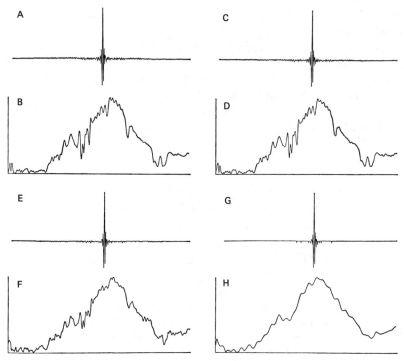

Fig. 2.6.   The effect of insufficient resolution of the ADC. The interferograms A, C, E and G were measured using a 13-bit, 8-bit, 6-bit and 4-bit ADC, respectively. The spectrum from a given interferogram is shown beneath. (Reproduced from [4] by permission of the author and the American Chemical Society; copyright © 1970.)

Perry et al. [5] have discussed digitization needs more quantitatively. They have approximated the shape of an emission background to a Gaussian curve, and have expressed the spectral intensity $S_b(\bar{v})$ at any frequency $\bar{v}$ as

$$S_b(\bar{v}) = S_b \exp\left[-\frac{2\sigma_b{}^2}{(\bar{v} - \bar{v}_b)^2}\right] \tag{2.4}$$

where S_b = peak amplitude
σ_b = bandwidth at half-height
$\bar{v}_b$ = center frequency.

The interferogram, $I_b(\delta)$, can be shown to be

$$I_b(\delta) = \left(\frac{\pi}{2}\right)^{\frac{1}{2}} S_b\sigma_b\{\exp\left[-(\pi\sigma_b\delta)^2\right]\} \cos(2\pi\bar{v}_b\delta). \tag{2.5}$$

For a Gaussian type absorption band superimposed on the background with a center frequency $\bar{v}_s$, peak amplitude S_s and half-width σ_s (Fig. 2.7), the corresponding interferogram can be written

$$I_s(\delta) = \left(\frac{\pi}{2}\right)^{\frac{1}{2}} S_s\sigma_s \left\{ \exp\left[-\frac{(\pi\sigma_s\delta)^2}{2} \right]\right\} \cos(2\pi\bar{v}_s\delta). \qquad (2.6)$$

In order to observe this interferogram $I_s(\delta)$ superimposed on the background interferogram, $I_b(\delta)$, the dynamic range of the digital output must be greater than $I_b(\delta)/I_s(\delta)$ at the zero retardation point. Consider the special case that the small band absorbs at the center frequency of the background, that is, $\bar{v}_b = \bar{v}_s$. Let the amplitude of the small spectral band be a factor α less than the amplitude of the background. For this special case, the dynamic range of the ADC must be greater than $\sigma_b/\alpha\sigma_s$. In the general case, the dynamic range must be greater than the ratio of the area of the background spectrum to the area of the absorption band. If quantitative intensity information is needed on the small band, the dynamic range should be at least 10 times larger than the ratio of the two areas.

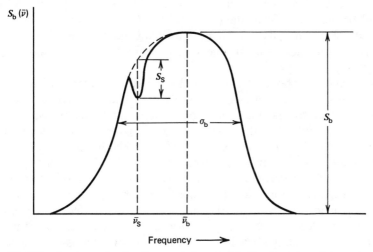

Fig. 2.7. Spectrum showing the parameters used in the discussion of dynamic range.

As the observation time per point increases, so does the signal-to-noise ratio in the sampled interferogram. Thus, in order to adjust the dynamic range of any interferogram to the level that it can be sampled correctly, the observation time must be changed until the two least significant bits of the ADC are sampling noise. The only way that the dynamic range can be reduced while maintaining a high observation efficiency is to increase the scan speed of the moving mirror. This is the reason why only rapid-scanning interferometers have been used for mid-infrared Fourier transform spectroscopy when a large incandescent source is being measured. It may also be noted that if only the last bits of the ADC are being used to digitize the noise level, a change to a more sensitive detector gives little gain in spectral signal-to-noise ratio under these conditions. On the other hand, slow-scanning interferometers can be used effectively for the measurement of weak sources often in conjunction with very sensitive detection systems.

There have been several attempts made to increase the effective dynamic range of digitizing systems. One of the most successful of these involves the use of dual-beam Fourier transform spectroscopy. Since this technique is discussed in greater detail in Chapter 7, it will be only briefly described here. Basically, the technique involves mixing the signals from both arms of the interferometer. Since these beams are exactly $180°$ out-of-phase, if the paths were identical they would destructively interfere to give no net signal. If an absorbing material were placed in one path between the beamsplitter and detector, the radiation absorbed by this sample would not be totally compensated and an interferogram caused only by the sample absorption bands would be measured. Thus the dynamic range of the interferogram is substantially less than if all the frequencies from the source were being measured. This is the equivalent of being able to measure $I_s(\delta)$ of Eq. 2.6 without having it superimposed on $I_b(\delta)$.

As mentioned above, the dynamic range of any interferogram can be reduced by scanning the moving mirror faster, and any signal can be sampled accurately (i.e., so that the least two or three significant bits of the ADC digitize noise) provided that the scan speed of the interferometer can be increased in order that the signal is reduced. The limiting factor for this method therefore becomes the response time of the detector, which can sometimes become a problem. For example, mid-infrared radiation is usually detected by thermal detectors (thermocouples, thermistor bolometers, etc.). However, in order to reduce the dynamic range of the interferogram measured from an incandescent source, the scan speed has to be increased to the point that the modulation frequencies of the shortest wavelengths in the spectrum are so high (>1 kHz) that their period is less than the response time of the detector (>1 msec). This presented a real problem in the measurement of mid-infrared spectra over a wide range until the advent of the pyroelectric bolometer. These detectors have a broad spectral response with a short response time (<1 msec) even though their noise-equivalent power is approximately equal to that of a thermistor bolometer.

Even when a pyroelectric bolometer (such as triglycine sulfate, TGS) is used for these measurements, the dynamic range of an interferogram from an incandescent source measured using a single scan of a rapid-scanning interferometer sometimes does not allow weak spectral features to be seen. For instance, for a spectrum with $\Delta\bar{v} = 2$ cm^{-1} and $\bar{v}_{max} = 4000$ cm^{-1} measured with a 14-bit ADC, any feature in this spectrum absorbing less than 0.25% of the radiation would be missed. To improve the signal-to-noise ratio achievable in the interferogram, signal-averaging techniques must be used. If this were the case, the final spectral signal-to-noise ratio would be decided not by the ADC but by the word-length of the signal averager. Thus, if a data system with a 16-bit word-length were used in single-precision for the signal-averaging process and computation, the ultimate dynamic range of the interferogram could be no larger than 16 bits and hence the spectrum would show bands absorbing 0.1% (but no less). To obtain a greater dynamic range in the spectrum, either double-precision techniques (i.e., using two words per data point) or a computer with a longer word-length would have to be used.

If the noise level in any waveform is much smaller than the resolution of the ADC then signal averaging will not improve the signal-to-noise of the signal-averaged waveform over the single-scan case, since the same value of any point will be measured for each scan. This leads to the odd situation that it is conceivably possible to obtain a greater spectral signal-to-noise ratio ultimately from an interferogram of small dynamic range than one of high dynamic range. It has been suggested that a random noise signal of limited bandwidth or a single, incoherent sinusoidal wave outside the frequency range of interest but below the folding frequency can be added to the interferogram. This raises the instantaneous signal so that it will be digitized accurately and signal averaging will allow the true value to be approached (Fig. 2.8).

Another technique used to increase the dynamic range of the interferogram is the use of a gain-ranging amplifier. In this method, an amplifier is used whose gain can be rapidly changed by a factor of two, four, eight, and so forth, a certain distance either side of zero retardation (Fig. 2.9). In this way, the region of less dynamic range away from zero retardation can be amplified and therefore sampled more accurately than if such an amplifier were not used. At the end of the measurement, the least significant bits of the signal-averaged interferogram in the region where the greater gain had been used are dropped from the interferogram. If the increase in gain is a factor of 2^N in the region of small dynamic range, then N bits would be dropped before the transform is performed.

Mark and Low [6] have suggested a technique for decreasing the dynamic range which involves "clipping" or "blanking" the interferogram around zero retardation. "Clipping" involves increasing the amplifier gain so that the region of largest amplitude has a greater intensity than the full dynamic range of the ADC. "Blanking" involves bringing the region around zero retardation to zero. These effects are illustrated in Fig. 2.10. In either case,

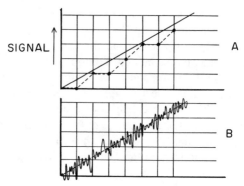

SIGNAL   A

B

Fig. 2.8. The effect of digitizing a signal with an ADC of limited dynamic range for signal-averaging systems. For a noise-free signal (a—solid line), the same value of the signal is measured each scan, and the apparent signal after one scan or many signal-averaged scans would be the same (broken line). However, if the peak-to-peak noise level is greater than the least significant bit (b—solid line), signal averaging takes place and the final signal (broken line) is a much closer approximation to a straight line than if no noise were present.

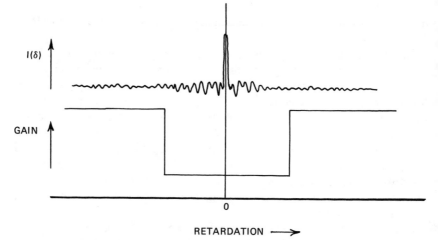

$I(\delta)$

GAIN

0

RETARDATION $\longrightarrow$

Fig. 2.9. The action of a gain-ranging amplifier.

70

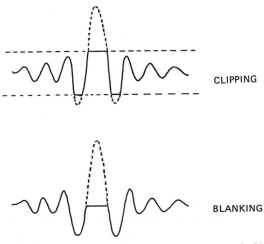

Fig. 2.10. (a) "Clipping" the interferogram, and (b) "Blanking" the interferogram around zero retardation.

a higher gain can be used so that the regions away from zero retardation (the "wings" of the interferogram) can be more accurately sampled. These techniques have the great disadvantage of putting a discontinuity in the interferogram, thereby adding a great deal of noise to all frequencies in the spectrum. Any benefit gained by sampling the wings of the interferogram is more than compensated by this increase in noise, and therefore the technique shows no practical purpose.

Another method of circumventing the problems of dynamic range has been used in the Infrared Interferometer Spectrometer (IRIS) experiment aboard the Nimbus B satellite and described by Forman [7]. All points that are less than 10% of the maximum are quantized to 256 levels (8-bit ADC). The remaining points are then divided by 10 and quantized to the same number of levels, and the operation is recorded by one bit of a 10-bit telemetry word. This system was designed because of the telemetry unit rather than the dynamic range of the ADC. Forman's data show not only the usefulness of this technique but also the effect of insufficient resolution of the ADC on computed spectra. His spectra are shown in Fig. 2.11.

In all the interferograms discussed to date there is a very large signal at one point (zero retardation) since each wavelength is in phase (or very close to it) at this point. It is possible to modify the interferogram such that each wavelength has its own origin, thus avoiding large central fringes that normally occur. This distribution of wavelength centers is accomplished by placing a flat optical element into each beam before the point of beam

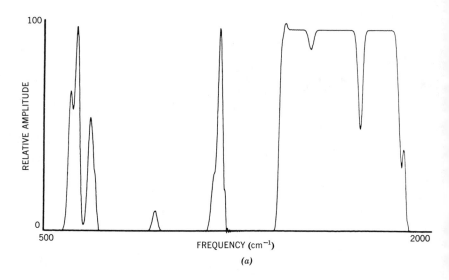

(a)

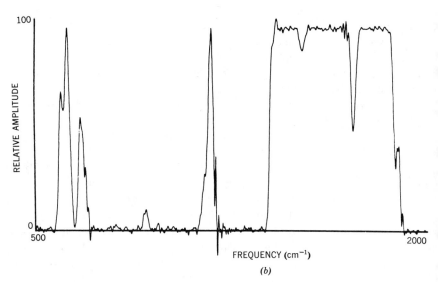

(b)

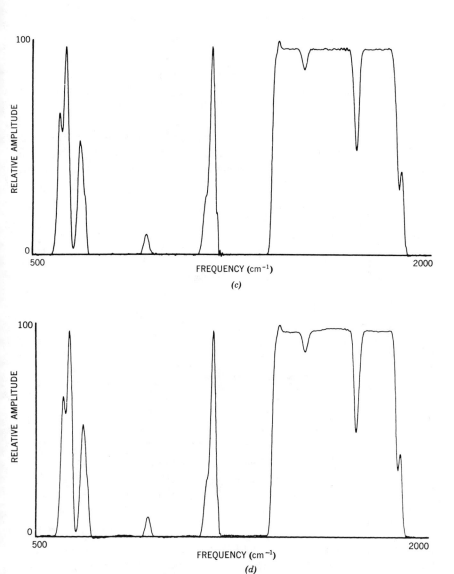

Fig. 2.11. (a) Simulated spectrum; (b) effect on the simulated data if the interferogram were sampled with an 8-bit ADC; (c) effect on the simulated data if the interferogram were sampled with a 10-bit ADC; and (d) effect on the simulated data if the interferogram were sampled using Forman's divide-by-ten method using an 8-bit ADC. (Reproduced from [7] by permission of the author.)

73

recombination. If the refractive index of one of the plates is frequency dependent or if the plates are of different thickness, then each wavelength will have its center shifted by a different amount, and the interferogram becomes strongly chirped. Therefore, there is no single path difference at which all wavelengths are in-phase to produce a central fringe, so that the dynamic range is reduced while retaining Fellgett's advantage.

Two materials that have been used for this purpose are Irtran 4 and Irtran 5. The displacement in the zero phase difference point for different frequencies (with CaF_2 in the other arm) is shown in Fig. 2.12. It can be seen that Irtran

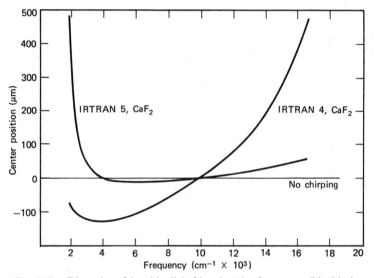

Fig. 2.12. Dispersion of the white-light fringe location for two possible chirping configurations. (Reproduced from [8] by permission of the author.)

5 spreads the zero *phase* difference fringe for the 2000 to 4000 cm^{-1} region over 500 μm of *path* difference. A PbSe detector has its peak response over this region and has been used in a "chirped" interferometer by Hohnstreiter et al. [8]. A typical interferogram is shown in Fig. 2.13. The interferogram shows that the low frequencies are spread over a wide range and the high frequencies converge to a single point. This is expected in the light of Fig. 2.12 since chirping is only expected to occur strongly below 4000 cm^{-1} for an Irtran 5, CaF_2 combination.

In the measurements described by Hohnstreiter et al. [8], the magnitude of the signal could vary by quite large amounts, and eight levels of automatic gain changing were required to keep the signal level within the telemetry

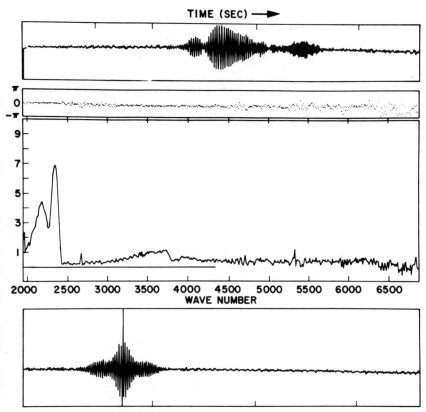

Fig. 2.13. (Above) Chirped interferogram and (center) the phase-corrected spectrum of the source which was calculated from this interferogram. (Below) An almost completely unchirped interferogram reconstructed from this spectrum. (Reproduced from [10] by permission of the author and the Society for Applied Spectroscopy; copyright ⓒ 1974.)

range. To obtain the maximum information from the interferogram under these conditions, it was found necessary to set the gain of the amplifier so that the maximum signal had a value small enough that it can be modulated for telemetry without clipping the interferogram, yet large enough that the principal source of noise was not the telemetry circuits. To achieve this objective, successive scans were made, and the gain was changed by factors of approximately four between scans if the maximum voltage from the previous interferogram exceeded a maximum, or was less than a minimum, threshold value. In this way the telemetered interferogram could be transmitted at the optimum gain. The complex data-handling equipment has been described by Sheahen et al. [9].

III. EFFECTS OF SAMPLING ERRORS

Along with the rapid development of the Michelson interferometer for Fourier transform spectroscopy, data systems have progressed to the stage that errors are rarely encountered during data collection. However, occasionally errors can still be generated and it is worthwhile to recognize their effects. The most important types of sampling error are the effects of missed or added sample points and the effect of "bad" points in the interferogram. Errors of the first type are rarely encountered but bad points can be generated mechanically by knocking the instrumentation during a measurement or electronically during the digitization or storage processes.

For both types of error the effect on the spectrum is dependent on how close to the zero retardation point the error occurs. A missed point in the interferogram of a monochromatic source results in a very rapid change of phase angle for the remaining points of the interferogram (Fig. 2.14). If the error occurs close to the zero retardation point, the resultant spectrum shows a broadening of the line width and a distortion of the line shape. This has been demonstrated by Horlick and Malmstadt [4] who have shown a series

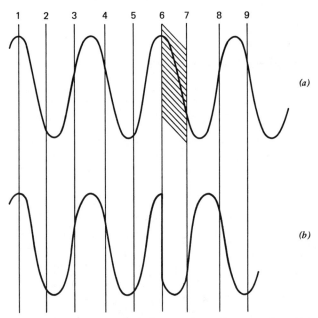

Fig. 2.14. The effect of missing one data point on a sinusoidal wave; the rapid phase change causes a decrease in frequency and photometric accuracy and an increase in the noise level of the spectrum.

of spectra of a He-Ne laser where a sample point has been missed at different retardations. It may be seen from Fig. 2.15 that the final spectrum (where a sample point was dropped near the end of the scan) is almost identical to the first spectrum where no point was missed or added. These spectra were computed using a triangular apodization function so that the final points in the scan have less weight in the computations than if a boxcar truncation function were used. In this latter case, the effect of a missed or added sample point at high retardation would be somewhat more prominent. Horlick and Malmstadt [4] have also shown the effect of an added point on the spectrum of an iron hollow cathode lamp; again the spectrum shows a decrease in

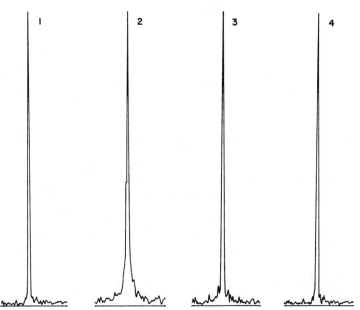

Fig. 2.15.   Effect of a missed sample point on the spectrum of a He-Ne laser. Spectrum #1 is the true spectrum; for spectra 2, 3, and 4, one sample point has been dropped from the interferogram at increasingly high retardation values. (Reproduced from [4] by permission of the author and the American Chemical Society; copyright © 1970.)

resolution and a slight increase in noise level (Fig. 2.16). A decrease in frequency accuracy cannot be seen from this spectrum but is found in practice.

The effect of a bad point in the interferogram is seen at every frequency in the spectrum. In the same way that the interferogram of a monochromatic source is a sinusoidal wave, if an incorrect sample point is measured in the *interferogram*, the transform shows a sinusoidal wave superimposed on all

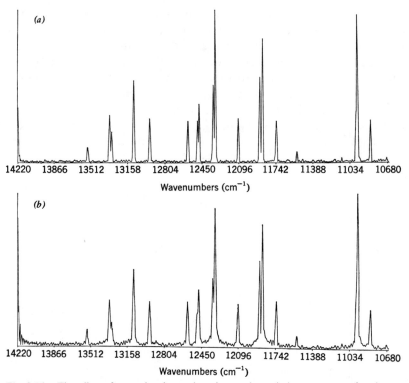

Fig. 2.16. The effect of one missed sample point on the emission spectrum of an iron hollow cathode lamp; the upper trace is the correct spectrum. It is seen that a slight increase in the noise level and a decrease in the resolution results. (Reproduced from [4] by permission of the authord and the American Chemical Society; copyright ©️ 1970.)

frequencies in the *spectrum*. The greater the retardation at which the bad point is generated, the higher the frequency of the sinusoid it generates. The larger the error in any sample point, the greater the amplitude of the sine wave in the spectrum.

Occasionally, sinusoidal waves from physical causes are seen in a spectrum. These causes include interference fringes from an empty cell or window in the beam or rotational fine structure in a far-infrared spectrum. These may be easily distinguished from the effect of a bad point since a bad point will generate a sine wave regardless of whether there is any real energy at a particularly frequency, whereas a sine wave from a real physical cause is seen only at frequencies where radiation is measured. Thus, merely by examining a region of the *single-beam* spectrum where no energy is expected (e.g., from 0 to 400 cm^{-1} where KBr optics are used for mid-infrared systems) it is simple to decide whether the sine wave seen in the spectrum is caused by a bad point or by a true physical phenomenon such as an interference fringe.

Another effect that can deteriorate spectral quality is a baseline of the interferogram which is not set at zero. It was stated in Chapter 1, Section II that only the a.c. portion of an interferogram is used in the computation of the spectrum. Thus, if the interferogram is measured by a slow-scanning interferometer, the average level of the interferogram must be accurately subtracted from each sample point. If this is not done, the resultant interferogram has a d.c. bias or the equivalent of an infinitely low frequency wave superimposed on it, so that a very large spike can be seen in the computed spectrum close to $0\,cm^{-1}$. Similarly, if the intensity of the source varies slowly, a large apparent signal will be seen in the spectrum at low frequency. The higher the frequency of the source variations, the higher the frequency in the spectrum at which the effects can still be seen. This source of error for slow-scan interferometers is one of the principal benefits derived from using phase modulation or a rapid-scanning interferometer for far-infrared spectroscopy, since the modulation frequency of all the spectral radiation of interest is much greater than the frequency at which the source intensity usually fluctuates.

One final source of error that can occur while sampling the interferogram involves the use of a gain-ranging amplifier. In Section II, the use of a gain-ranging amplifier to amplify regions away from zero retardation and hence increase the effective dynamic range was discussed. If the gain change is not exactly 2^N (where N is an integer) then when the amplified regions of the interferogram are divided by (exactly) 2^N, the effective gain of this region will be slightly different from the rest of the interferogram. This effect will result in a sinc function centered around $0\,cm^{-1}$ being seen in the spectrum (Fig. 2.17).

It should be stressed that these errors are rarely encountered in practical Fourier transform spectroscopy, but it certainly helps the practicing spectroscopist to recognize an instrumental problem if he can correlate an unusual feature in a spectrum with a possible cause.

REFERENCES

1. M. Woodward, *Probability and Information Theory*, Pergamon Press, New York, 1955.
2. P. R. Griffiths, C. T. Foskett, and R. Curbelo, *Appl. Spectrosc. Revs.*, **6**, 31 (1972).
3. P. R. Griffiths, C. T. Foskett, R. Curbelo, and S. T. Dunn, *Inst. Soc. Am.*, *Analysis Instrumentation*, **8**, II–4 (1970).
4. G. Horlick and H. V. Malmstadt, *Anal. Chem.*, **42**, 1361 (1970).
5. C. H. Perry, R. Geick, and E. F. Young, *Appl. Optics*, **5**, 1171 (1966).
6. H. Mark and M. J. D. Low, *Appl. Spectrosc.*, **25**, 605 (1971).

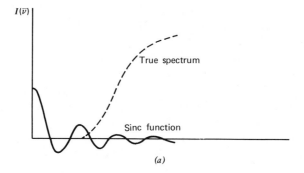

$I(\bar{\nu})$

True spectrum

Sinc function

(a)

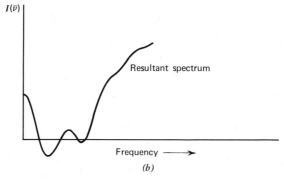

$I(\bar{\nu})$

Resultant spectrum

Frequency ⟶

(b)

Fig. 2.17. The effect of a badly adjusted gain-ranging amplifier: (a) shows the true spectrum which would have been measured had the amplifier been in good adjustment (broken line), together with the sinc function which results from transforming the boxcar function due to the maladjusted amplifier (solid line); (b) shows the resultant spectrum.

7. M. L. Forman, Aspen Int. Conf. on Fourier Spectrosc., 1970, G. A. Vanasse, A. T. Stair, and D. J. Baker Eds., AFCRL–71–0019, p. 305.
8. G. F. Hohnstreiter, W. Howell, and T. P. Sheahen, Aspen Int. Conf. on Fourier Spectrosc., 1970, G. A. Vanasse, A.T. Stair, and D. J. Baker Eds., AFCRL–71–0019, p. 243.
9. T. P. Sheahen, G. F. Hohnstreiter, W. R. Howell, and I. Coleman, ibid., p. 255.
10. T. P. Sheahen, *Appl. Spectrosc.*, **28**, 283 (1974).

COMPUTING TECHNIQUES

I. THE CONVENTIONAL FOURIER TRANSFORM

In the two previous chapters, the various factors for the generation of an interferogram were discussed. In this chapter the techniques for computing the spectrum from this digitized interferogram are described.

Even as early as 1892, the importance of a large data-handling capability was realized by Michelson [1]. The ingenious analog computer designed by Michelson and described in the Introduction was the first attempt to compute spectral information from the output of an interferometer. However it was over half a century later that the first successful computation of a spectrum from an interferogram was performed by Fellgett [2]. Not unnaturally, the time at which this calculation was performed corresponded with the time at which digital computers were first becoming available.

Between 1950 and 1966, spectroscopists who measured spectra interferometrically used the same basic algorithm for their computations. This involved the use of what is now known as the *conventional* or *classical* Fourier transform. Although it is true that few people today use this algorithm in view of the substantial time advantages to be gained by the use of the fast Fourier transform technique (which will be described in the next section), an understanding of the conventional Fourier transform leads to a better comprehension of more advanced techniques.

Let us first consider [3] the case of a symmetrical interferogram that has been measured from the zero retardation point. The integral

$$B(\bar{v}) = \int_{-\infty}^{+\infty} I(\delta) A(\delta) \cos 2\pi \bar{v} \delta \cdot d\delta, \qquad (3.1)$$

where $A(\delta)$ represents any truncation or apodization function previously referred to as $D(\delta)$ or $A_1(\delta)$, $A_2(\delta)$, and so forth, can be written as

$$B(\bar{v}) = 2 \int_{0}^{+\infty} I(\delta) A(\delta) \cos 2\pi \bar{v} \delta \cdot d\delta \qquad (3.2)$$

because of the symmetry of the interferogram about the point of zero retardation. It was stated earlier that if the signal was digitized with a small enough interval between data points, no information from the analog signal would be lost. The minimum sampling interval, h, is given by

$$2h = (\bar{v}_{max} - \bar{v}_{min})^{-1}. \tag{3.3}$$

If the interferogram has been sampled correctly we can replace the integral of Eq. 3.2 by a summation, and calculate the intensity $B'(\bar{v}_1)$ at any frequency $\bar{v}_1$. Let the retardation at any sampling point be given by nh, where n is an integer, and let $I_a(n)$ be the value of the interferogram at this point:

$$B'(\bar{v}_1) = h\{I_a(0) + 2I_a(1) \cos(2\pi\bar{v}_1 \cdot h) + 2I_a(2) \cos(2\pi\bar{v}_1 \cdot 2h) \cdots$$
$$+ 2I_a(N - 1) \cos[2\pi\bar{v}_1 \cdot (N - 1)h]\} \tag{3.4}$$

where N is the total number of points sampled.

$$B'(v_1) = h\left[I_a(0) + \sum_{k=1}^{(N-1)} 2I_a(k) \cos(2\pi\bar{v}_1 \cdot kh)\right] \tag{3.5}$$

The summation shown in Eq. 3.5 is performed for all frequencies of interest in the spectrum. It need not, however, be carried out for all frequencies from $\bar{v}_{max}$ to $\bar{v}_{min}$.

Even for fairly short spectral ranges, the computing time involved for this operation can be quite high especially if each cosine term required is calculated from a series. Cosine tables could be stored in the computer memory, but for large interferograms this would take a large amount of space. It has been found that a successful way of circumventing this problem is by the use of recurrence relationships, whereby a value for $\cos(2\pi\bar{v}_1 \cdot kh)$ can be calculated from the value of $\cos(2\pi\bar{v}_1 h)$. In this way, only one cosine value has to be calculated for each spectral frequency. The Chebyshev formula,

$$\cos(p + 1)x = 2 \cos x \cos px - \cos(p - 1)x, \tag{3.6}$$

is the one used most frequently. It has been shown that if the cosines are calculated to six significant figures, the difference between the value computed by a series expansion and the value obtained after 12,000 iterations is less than one in 50,000. For each new value of $\bar{v}_1$ the first cosine, $\cos 2\pi\bar{v}_1 h$, is computed using series and all the cosines that follow are computed using a recurrence formula. Each of the values of $I_a(k) \cos(2\pi\bar{v}_1 \cdot kh)$ is added to the sum and the resultant number is proportional to the value of $B'(\bar{v}_1)$.

If a double-sided transform is desired, the summation given in Eq. 3.5

must be replaced by the summation:

$$B'(\bar{v}_1) = \left\{ \left[\sum_{k=-\frac{N}{2}}^{(\frac{N}{2})-1} I_a(k) \cos (2\pi\bar{v}_1 \cdot kh) \right]^2 \right.$$

$$\left. + \left[\sum_{k=-\frac{N}{2}}^{(\frac{N}{2})-1} I_a(k) \sin (2\pi\bar{v}_1 \cdot kh) \right]^2 \right\}^{\frac{1}{2}}. \quad (3.7)$$

The sine values are calculated using another recurrence formula also attributed to Chebyshev:

$$\sin (p + 1)x = 2 \sin px \cos x - \sin (p - 1)x \quad (3.8)$$

If there are N points in the interferogram, let us consider how many operations are required during the computation of a single-sided transform. For each spectral frequency, $\bar{v}_j$, we perform N multiplications for $I_a(k) \cos (2\pi\bar{v}_j \cdot kh)$. To add the values we must do N additions, so that a total of $2N$ operations are needed for each frequency $\bar{v}_j$. To examine the complete spectrum we must compute the spectrum at $N/2$ frequency values at least. Thus a total of N^2 operations are needed for the conventional Fourier transform, not counting the time required to compute the cosine value.

If the double-sided transform is performed, the number of operations increases. For each value of $\bar{v}_j$, we now perform N multiplications for $I_a(k) \cos (2\pi\bar{v}_j \cdot kh)$ and N multiplications are required for $I_a(k) \sin (2\pi\bar{v}_j \cdot kh)$. N additions must be made for the cosine transform and N for the sine transform, so that a total of $4N$ operations must be made for each frequency. If the full bandwidth of the spectrum is to be examined at the same resolution as the single-sided spectrum, twice as many data points must be taken to achieve the same resolution, so that the number of operations is increased in practice to $8N$. The number of output points is the same as for the single-side case, so that $4N^2$ operations are required. This number does not include the time required for the cosine, sine, square, and square root calculations required for each value of $\bar{v}_j$.

As the number of points to be calculated increases beyond about 10,000 the calculation time for a spectrum becomes prohibitive. In view of the fact that interferometry shows its greatest advantage over dispersive spectroscopy at high resolution, that is, when large numbers of points have to be calculated, this problem was especially annoying and did not appear to be resolvable until about 1966. At this time Forman [4] published a paper on the application of the fast Fourier transform technique to Fourier spectroscopy. This technique had been described in the literature by Cooley and Tukey [5] one year earlier. This algorithm extended the use of Fourier transform spectroscopy to encompass high resolution data in all regions of the infrared spectrum. It is described in the next section.

II. THE FAST FOURIER TRANSFORM (FFT)

To best understand the use of the FFT for interferometry, the interferogram and the spectrum should be regarded as the complex pair [6]:

$$I(\delta) = \int_{-\infty}^{+\infty} B(\bar{v}) e^{2\pi i \bar{v} \delta} \cdot d\bar{v} \tag{3.9}$$

$$B(\bar{v}) = \int_{-\infty}^{+\infty} I(\delta) e^{-2\pi i \bar{v} \delta} \cdot d\delta \tag{3.10}$$

where Eq. 3.9 is an alternative way of writing Eq. 1.23.

The theory of the FFT will be discussed [7, 8] with respect to an arbitrary complex function, I_k, which has been sampled at $2N$ discrete intervals, so that $k = 0, 1, 2, \ldots, 2N - 1$. For interferometry, I_k would be the truncated or apodized interferogram. Let B_r be the discrete Fourier transform of I_k:

$$B_r = \sum_{k=0}^{2N-1} I_k e^{2\pi i r k / 2N}. \tag{3.11}$$

(B_r, therefore, corresponds to $B'(\bar{v}_j)$ in the previous section, and r corresponds to the output frequencies $\bar{v}_j$.)

Let us define a parameter W, such that

$$W = e^{2\pi i / 2N} = e^{\pi i / N} \tag{3.12}$$

Substituting this value of W in Eq. 3.11, we obtain

$$B_r = \sum_{k=0}^{2N-1} I_k W^{rk} \tag{3.13}$$

The most important rule for the computation of B_r from I_k using the Cooley-Tukey algorithm rests on the fact that the discrete Fourier transform (DFT) of any given complex number is a linear combination of the two DFT's of two functions with half as many terms and issued from the first.

Let us break up the set of I_k into those with even k indices, which we label Y_k, and those with odd k indices, which we label Z_k.

$$Y_k = I_{2k}$$
$$Z_k = I_{(2k+1)}$$

with $k = 0, 1, 2, \ldots, N - 1$.

Let C_r and D_r be the DFT of Y_k and Z_k, respectively, each having only N terms.

$$C_r = \text{DFT}(Y_k) = \sum_{k=0}^{N-1} Y_k \, e^{2\pi i r k/N} = \sum_{k=0}^{N-1} Y_k \, W^{2rk} \qquad (3.14)$$

$$D_r = \text{DFT}(Z_k) = \sum_{k=0}^{N-1} Z_k \, W^{2rk} \qquad (3.15)$$

with $r = 0, 1, 2, \ldots, N - 1$.

If we separate the even and odd terms in Eq. 3.13, we get

$$B_r = \sum_{k=0}^{N-1} Y_k \, W^{2rk} + \sum_{k=0}^{N-1} Z_k \, W^{(2k+1)r} \qquad (3.16)$$

so that

$$B_r = C_r + W^r D_r. \qquad (3.17)$$

The functions C_r and D_r are periodic, with an equal period, N; from this fact it is possible to show that

$$C_{(N+r)} = C_r \qquad (3.18)$$

$$D_{(N+r)} = C_r \qquad (3.19)$$

$$W^{(r+N)} = -W^r \qquad (3.20)$$

We can obtain the two sets of samples B from 0 to $N - 1$ and from N to $2N - 1$ with the two linear combinations:

$$B_r = C_r + W^r D_r \qquad (3.17)$$

and

$$B_{N+r} = C_r - W^r D_r \qquad (3.21)$$

with $r = 0, 1, 2, \ldots, N - 1$.

Thus, we can compute the $2N$ complex samples of the DFT of the entire interferogram I_k from the N complex samples of each of the two subjects Y_k and Z_k. If we apply the same reasoning $\log_2 2N$ times (provided that N is a power of two), we can obtain B_r, the discrete Fourier transform of I_k with only $2N \log_2 2N$ elementary operations.

Connes [7] has shown this diagrammatically for the case with $2N = 8$. She uses the symbol P for the discrete Fourier transform operation. In Fig. 3.1a, the eight values of I_k have been sorted into two subsets P_1 and P_2 corresponding to the even and odd indices of I_k. To each of these subsets, a DFT is applied, and the operations described by Eqs. 3.17 and 3.21 give the final eight output points from the two partial DFT's.

A new subdividing of P_1 into P_3 and P_4 and P_2 into P_5 and P_6 again gives the final eight values of B_r, but the operations in Eqs. 3.17 and 3.21 have to be applied twice (Fig. 3.1c). One more step in which P_3, P_4, P_5, and P_6 are further subdivided is shown in Fig. 3.1d in which are presented all the elementary operations of the FFT in the case of $2N = 8$.

The repetitive division into odd and even groups eliminates many unnecessary additions and multiplications for the computer program. This is the basis of the time saving gained by the use of the Cooley-Tukey algorithm over the conventional Fourier transform and accounts for its other name, the "decimation in time" method.

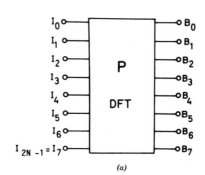

(a)

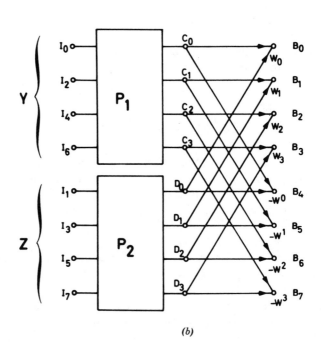

(b)

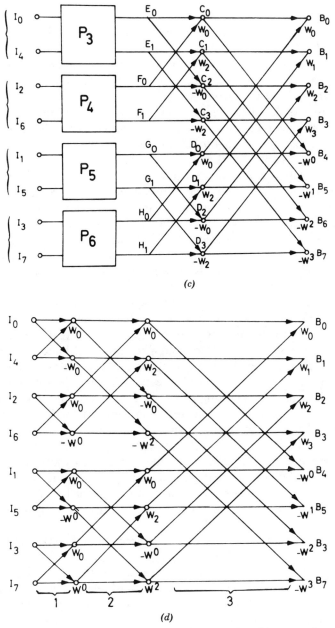

(c)

(d)

Fig. 3.1. (a) Discrete Fourier transform of 8 input points; (b) computation of the sampled spectrum after separating Set P into Subsets P_1 and P_2; (c) separation of Sets P_1 and P_2 into P_3, P_4, P_5, and P_6; (d) complete diagram of the "Decimation-in-Time" FFT technique. (Reproduced from [7] by permission of the author.)

In the general case of the fast Fourier transform of an arbitrary complex function, the following procedures are carried out: (a) division of the subsets given by the previous operation into two parts; (b) combination of the two results of the DFT operation applied to the new subsets in order to recover the DFT results of the previous stage.

There are $\log_2 2N$ stages involved in (a), while the combination described in (b) implies one multiplication and one addition. Hence the total number of operations is proportional to $2N \log_2 2N$.

The derivation above applies to an arbitrary complex function. However, symmetric interferograms are real functions; thus, if we use the technique outlined above, we are wasting effort and computing space, since we know that the imaginary part of the interferogram, I_k, is zero for all k. In the general case, half the computer locations (the ones containing the imaginary part of the interferogram) would therefore be filled with zeros. It is possible to save this space and use $2N$ rather than $4N$ locations for the computation. In this way the number of operations is reduced by a factor of approximately two from the case for a complex function.

It is further known that these symmetric interferograms are not only real functions but also even functions (due to their symmetry about the zero retardation point). By applying certain properties of real even functions, Connes [7] has shown that computer space can be reduced by a factor of four over the case of an arbitrary complex function and by a factor of two over an arbitrary real function. When extremely large transforms are being performed, these savings in computer space can be very important, since even third generation computers have fairly small amounts of directly accessible memory. When auxilary memory which is not directly accessible has to be used, computing time can go up drastically.

In view of additional operations besides those directly attributable to the FFT, the *time* savings in these computations are not quite as great as the *space* savings. Connes has reduced the computing time by a factor of 1.5 by considering the interferogram to be an arbitrary real function and by 2.8 by considering it as a real, even function.

If phase modulation is used in the measurement of the interferogram, the interferogram is no longer a real, even function but a real, odd function (see Chapter 1, Section XI. Although the mathematics of the transform are formally different from the normal case, equal time and space savings can result in the computation of the spectrum from a phase modulated interferogram using the FFT algorithm.

If the transform time for the case of the FFT of an arbitrary complex function is compared to that for a conventional Fourier transform of a double-sided interferogram, the time saving is seen to be $2N/\log_2 2N$. The practical time saving for the FFT of a single-sided interferogram is quite similar.

To visualize the importance of these time-saving factors for chemical spectroscopists, consider the case of a standard mid-infrared reference spectrum. The Coblentz Society [9] has suggested that these should be measured from 3800 to 450 cm^{-1} at a resolution of 2 cm^{-1}. This means that if the measurement were performed interferometrically, $8K$ data points would have to be collected for both the sample and reference interferograms. The time saving for a 8192 point interferogram if transformed using the FFT is approximately a factor of 400 over the conventional Fourier transform. The computing time for two $8K$ point interferograms is under two minutes, even on a mini-computer with a nondirectly accessible memory, while with a conventional transform, the operation would have taken up to a quarter of an hour using the best computer available. Since a reference spectrum can be measured on a good grating spectrometer in a little over half an hour, the time advantage of the interferometer using a conventional Fourier transform becomes negligible, even though the interferograms can be measured in a matter of seconds.

III. ZERO-FILLING

As described above, the complex FFT produces one output point in the real array and one output point in the imaginary array for each two input (interferogram) points, so that only one real (spectrum) point is calculated per resolution element. Although it is theoretically true that this number of points is sufficient to define the spectrum, a linear interpolation between the output points results in an obvious stepping between each point. A direct result of this stepping effect is a loss in the photometric precision of Fourier transform spectroscopy.

When the complex FFT of N interferogram points is performed, the real output array contains $N/2$ points and the imaginary array also contains $N/2$ points. If N zeros are added to the input array, the real and imaginary output arrays then each contain N points, of which $N/2$ are linearly independent and the rest represent sinc x interpolations between these points. Although a smoother plot results from calculating more spectrum points in this fashion, distinct breaks are still seen between each data point, as evidenced by the portion of the spectrum of carbon monoxide shown in Fig. 3.2a, which was calculated after adding $8K$ zeros to an $8K$ point interferogram.

In the general case, if $(2^m - 1)N$ zeros are added to an N point interferogram, where m is an integer greater than one, the real array after the complex FFT will contain $2^{(m-1)}N$ points, of which $N/2$ are linearly independent and the rest are interpolated. This procedure, which is known as *zero-filling* [10], results in a far smoother spectrum, as demonstrated in Fig. 3.2. Figure 3.2a shows a spectrum of carbon monoxide computed with $m = 1$, while Fig. 3.2b shows the same interferogram computed with $m = 3$.

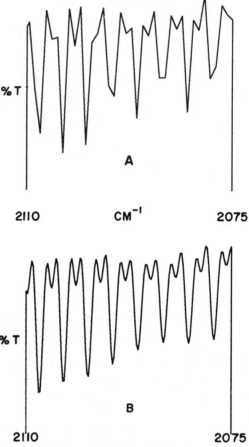

Fig. 3.2. Spectra of carbon monoxide from 2110 to 2075 cm^{-1} showing the effect of zero-filling; spectra were measured at an effective resolution of 2 cm^{-1} with boxcar truncation. (a) $m = 1$; 1 independent and 1 interpolated point per resolution element; (b) $m = 3$; 1 independent and 7 interpolated points per resolution element. (Reproduced from [10] by permission of the Society for Applied Spectroscopy; copyright © 1975.)

$m = 1$ represents the minimum amount of zero-filling which should be carried out on any interferogram that is not symmetrical about the zero retardation point and for which a complex FFT must be performed. However, for really good photometric precision, at least *eight* output points per resolution element are necessary, for which $m = 3$. Zero-filling obviously increases the computation time considerably, so that there is a trade-off between computing time and photometric precision. In addition, when data systems with limited storage are being used for the computations, it may not be possible to perform zero-filling to any great extent.

When a zero-filled interferogram of, say, carbon monoxide is transformed after multiplying the complete array with a triangular apodization function, it is found that the width of the lines becomes reduced below that of the non-zero-filled triangularly apodized case, and side-lobes are generated for each spectral line of similar magnitude to those seen in Fig. 3.2b. The reason for this effect can be illustrated by considering what happens when you add $3N$ zeros to an N point interferogram. If the resultant $4N$ point data array is triangularly apodized, the result is equivalent to multiplying a $4N$ point non-zero-filled interferogram by an apodizing function of the type shown in Fig. 3.3a. Any time that there is a sharp break in the apodization function, such as this case or the case of boxcar truncation, the ILS always shows important side-lobes.

To compute a spectrum from a zero-filled interferogram so that it has the same resolution and ILS as the spectrum computed from a triangularly apodized, non-zero-filled interferogram, an apodization function of the type shown in Fig. 3.3b should be used. This particular function is illustrated for a $4K$ interferogram, zero-filled with $12K$ zeros ($m = 2$).

It should be noted at this point that whereas zero-filling is the fastest method of computing interpolated data points over the *complete* spectrum, other methods of adding interpolated data points can also be used. In particular, a polynomial of two or three terms can be fitted to three data points very rapidly and interpolations between small numbers of data points can be made more quickly using this method than by the zero-filling technique.

IV.   REAL-TIME FOURIER TRANSFORMS

The techniques that have been described in the previous two sections can only be used to compute the spectrum after the end of the measurement of the interferogram. It is occasionally useful to compute the spectrum *during* the measurement of the interferogram in order to determine, for example, if the noise level in the spectrum is too great to justify any further increase in the retardation, or to check if the spectral information is increasing with increasing retardation. The techniques to be described in this section can only be used if the data rate is sufficiently low that the computations for one data point are finished before the next data point is measured. Thus spectra can rarely be computed from rapid-scanning interferometers using this technique, especially in view of the fact that signal-averaging is commonly

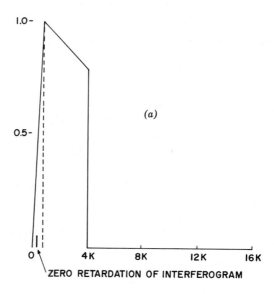

(a)

ZERO RETARDATION OF INTERFEROGRAM

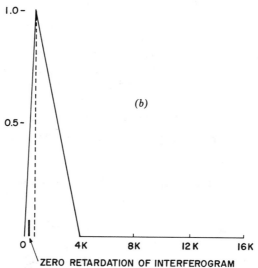

(b)

ZERO RETARDATION OF INTERFEROGRAM

Fig. 3.3. (a) Effective apodization function if a triangular apodization function is applied to a complete zero-filled array; (b) apodization function where only the *measured* interferogram points are weighted; when zero-filling is being carried out, this type of apodization function must be used to eliminate side-lobes. (Reproduced from [10] by permission of the Society for Applied Spectroscopy; copyright © 1975.)

92

used to improve the signal-to-noise ratio in the interferogram before the transform.

The techniques that have been developed for the purpose of computing the spectrum during the measurement of the interferogram are generally known as "real-time" Fourier transforms. It is generally only possible to examine a small spectral range (1000 output points is a typical number) but this can make it possible to examine a complete far-infrared spectrum at medium resolution or part of a high-resolution spectrum at any frequency.

The conventional Fourier transform involves several sums for each frequency value:

$$B(\bar{v}_1) = \tfrac{1}{2} I_a(0) + I_a(1) \cos (2\pi\bar{v}_1 \cdot h) + I_a(2) \cos (2\pi\bar{v}_1 \cdot 2h) + \cdots$$
$$B(\bar{v}_2) = \tfrac{1}{2} I_a(0) + I_a(1) \cos (2\pi\bar{v}_2 \cdot h) + I_a(2) \cos (2\pi\bar{v}_2 \cdot 2h) + \cdots$$
$$B(\bar{v}_i) = \tfrac{1}{2} I_a(0) + I_a(1) \cos (2\pi\bar{v}_i \cdot h) + I_a(2) \cos (2\pi\bar{v}_i \cdot 2h) + \cdots.$$

$$(3.22)$$

The usual calculation is made in each row after scanning of the interferogram. For real-time transforms [11], the calculation is made in each column, that is, in parallel for each frequency $\bar{v}_i$. The cosines $\cos 2\pi\bar{v}_i \cdot kh$ must therefore be calculated at each sampling point or taken from a stored cosine table, if necessary using interpolation routines. Recurrence relationships cannot be used in this case.

Let the number of frequency points in the output be M, so that the spectral intensity is being computed at frequencies $\bar{v}_1, \bar{v}_2, \bar{v}_3, \ldots, \bar{v}_i, \ldots, \bar{v}_M$. After k input points have been measured, M computer locations corresponding to the frequencies 1 through M contain the sum $B_{(k-1)}(\bar{v}_i)$, where

$$B_{(k-1)}(\bar{v}_i) = \tfrac{1}{2} I_a(0) + \sum_{j=1}^{(k-1)} I_a(j) \cos (2\pi\bar{v}_i \cdot jh).$$

$$(3.23)$$

When the $(k + 1)$th point is taken, the product $I_a(k) \cos (2\pi\bar{v}_i \, kh)$ is calculated and added to each value of $B_{(k-1)}(\bar{v}_i)$, from $i = 1, 2, \ldots, M$, currently in the memory, to give $B_k(\bar{v}_i)$. Each value of $B_k(\bar{v}_i)$ replaces the value of $B_{(k-1)}(\bar{v}_i)$ in the computer memory. The current values of the M locations can be viewed on an oscilloscope to visually check how the spectrum is changing with retardation, and the interferogram can be simultaneously recorded for computation of the complete measurement at the end of the measurement.

In practice, "real-time" computing techniques have fairly limited application to *chemical* Fourier transform spectroscopy. One reason for this is the high data rate required for reasons of dynamic range when an intense source is being measured in the mid infrared. This type of measurement necessitates the use of signal averaging for the improvement of signal-to-noise ratio, so that the use of real-time computing of the spectrum presents no positive benefits. Another reason is that even if there is as much as a one second interval between data points, the number of output points is still limited since the equivalent of a Cooley-Tukey FFT cannot be performed in the "real-time" mode. Thus, for mid-infrared chemical spectroscopy where relatively large numbers of data points are collected per interferogram at a high data rate using signal-averaging techniques, the optimum computing technique is usually an FFT at the completion of data collection.

For low resolution far-infrared spectroscopy, the complete spectrum can be computed between input data points provided that a slow-scanning interferometer is used for the measurement. In this case, however, there is the possibility of two sources of error. First, it was noted in Chapter 1, Section II that an interferogram measured in this way consists of an a.c. and a d.c. component. The first step in any Fourier transform calculation is to determine the mean value of the interferogram and subtract this value from each measured data point. This is most important for computing spectra using real-time techniques, and therefore the mean value of the interferogram must be determined before the measurement is started. Levy et al. [12] have performed this subtraction electronically, backing off the signal at the infrared detector by a stable constant voltage so that the average value of the interferogram well away from zero retardation is set as precisely as possible at zero volts. The effect of a drifting of the total signal measured at the detector during the measurement would be the introduction of a large amount of false energy at low frequency. (This effect is not an effect of the real-time technique, and would be equally noted if the transform were carried out at the end of the measurement. The only way to compensate for such a drift is by the use of a rapid-scanning interferometer whose band-pass is outside the frequency of the source drift, or by phase modulation.)

The second source of error involves the determination of the zero retardation point. Any slight sampling error in the position at which the first data point is taken means that the instrumental line shape will be distorted, as discussed in Chapter 1, Section VI. To get the best results with real-time computing techniques, the sampling error, ε, should be calculated by taking two or three points on either side of zero retardation, fitting a parabola to these points, and determining the retardation at which the maximum signal occurs. The sampling error, ε, can then be computed and applied to the summation shown in Eq. 3.23 to give:

$$B_{(k-1)}(\bar{v}_i) = \tfrac{1}{2} I_a(\varepsilon) + \sum_{j=1}^{(k-1)} I_a(j) \cos \left[2\pi\bar{v}_i(jh + \varepsilon)\right] \tag{3.24}$$

If the interferogram is not symmetric about the zero retardation point, it may be deduced that the phase angle $\theta_{\bar{v}}$ is finite, and that it probably varies with frequency. In this case the more sophisticated phase correction routines described in Chapter 1, Section VI must be applied. Since these involve calculations on a relatively large (20–100 points) region of the interferogram on both sides of zero retardation, "real-time" computations could not be started until after the first points in the interferogram had been calculated, which would be an extremely difficult computing problem unless the mirror drive were stopped at this point. Thus, while the technique could be used in this case, it can really only be used effectively in conjunction with an interferometer with a stepping-motor drive.

REFERENCES

1. A. A. Michelson, *Phil. Mag.* Ser. 5, **34**, 280 (1892).
2. P. B. Fellgett, Ph.D. thesis, University of Cambridge (1951).
3. J. Connes, *Rev. Opt.*, **40**, 45, 116, 171, 233 (1961). English translation as Document AD 409869, Clearinghouse for Federal Scientific and Technical Information, Cameron Station, Va.
4. M. L. Forman, *J. Opt. Soc. Am.*, **56**, 978 (1966).
5. J. W. Cooley and J. W. Tukey, *Math. Comput.*, **19**, 297 (1965).
6. D. C. Champeney, "Fourier Transforms and their Physical Applications," in G. K. T. Conn and K. R. Coleman, Eds., *Techniques of Physics, Vol. 1*, Academic Press, London, 1973.
7. J. Connes, Aspen Int. Conf. on Fourier Spectrosc., 1970, G. A. Vanasse, A. T. Stair, and D. J. Baker Eds., AFCRL–71–0019, p. 83.
8. B. Gold and C. Rader, *Digital Processing of Signals*, McGraw-Hill, New York, 1969.
9. Coblentz Society Board of Management, *Anal. Chem.* **38**, 27A (1966).
10. P. R. Griffiths, *Appl. Spectrosc.*, **29**, 11 (1975).
11. H. Yoshinaga et al., *Appl. Opt.* **5**, 1159 (1966).
12. F. Levy, R. C. Milward, S. Bras, and R. LeToullec, Aspen Int. Conf. on Fourier Spectrosc., 1970, G. A. Vanasse, A. T. Stair, and D. J. Baker, Eds., AFCRL–71–0019, p. 331.

4

TWO-BEAM INTERFEROMETERS

I. DRIVE SYSTEMS FOR MICHELSON INTERFEROMETERS

A. INTRODUCTION

The simplest version of the Michelson interferometer consists of two plane mirrors, one of which can move in a direction perpendicular to the plane of the other, with a beamsplitter between them. The interferometer may be considered as consisting of two separate parts, the drive mechanism for the moving mirror (including the sampling trigger) which will be described in the first section of this chapter, and the beamsplitter which will be considered in the second section.

It was noted in Chapter 1, Section IX that the quality of the drive mechanism must be higher

(a) the higher the desired resolution, and

(b) the higher the maximum frequency in the spectrum.

Thus, it is not surprising to find that the crudest drive mechanisms to have been successfully used for infrared spectroscopy were developed for far-infrared interferometers designed for medium and low resolution ($\geqslant 0.1 \text{ cm}^{-1}$) spectroscopy. The quality of the drive for high resolution far-infrared spectroscopy and medium resolution mid- and near-infrared spectroscopy must be higher; however such drive mechanisms have still been constructed at a moderate cost. When ultra high resolution ($\leqslant 0.01 \text{ cm}^{-1}$) is required in the near infrared, great care has to be taken in the design of the drive mechanism for the desired resolution to be attained, and only a very few such interferometers have been constructed.

Interferometers will be described for each of these categories, starting with systems designed for medium resolution far-infrared spectroscopy and ending with a description of very high resolution interferometers where corner and cats-eye retroreflectors are used to compensate for tilting errors that occur during a scan.

B. MEDIUM RESOLUTION FAR-INFRARED INTERFEROMETERS

Interferometers designed specifically for far-infrared spectroscopy at low or medium resolution have all been of the slow-scanning variety. The com-

96

mercial instrumentation available for this purpose nicely illustrates the simplicity of the drive required for this region, and several workers have constructed their own interferometers for far infrared spectroscopy. Far-infrared interferometers can use either a slow continuous scan or the step-and-integrate method. The continuous scanning instruments generally have an associated Moiré fringe reference device so that the interferogram can be sampled at equal intervals of retardation. Interferometers which have stepping motors do not require any such fringe referencing for far-infrared measurements, since the accuracy of the stepping motors is sufficiently high that the retardation may be increased by equal increments without adding any noise to the signal, (see Fig. 8.4.) For continuous scanning interferometers, the scan speed is usually variable, generally from about 0.5 μm/sec to 500 μm/sec, to allow the signal-to-noise ratio in the interferogram to be controlled; as the scan speed is decreased, the signal-to-noise ratio of the signal increases. For the step-and-integrate drive mechanisms, the time during which the movable mirror is held at each position is variable from 0.5 to 5 sec and occasionally more. The specifications of the instruments available from the major manufacturers are summarized in Table 4.1

TABLE 4.1

Manufacturer	Model	Drive Type	Maximum Retardation	Maximum Frequency
Beckman-RIIC [1]	FS–720	Continuous	10 cm	500 cm^{-1}
Beckman-RIIC	FS–820	Stepping	5 cm	500 cm^{-1}
Grubb-Parsons [2]	IS–3	Stepping	10 cm	675 cm^{-1}
Grubb-Parsons	Mark II	Stepping	2 cm (10 cm optional)	200 cm^{-1}
Coderg [3]	Fourier-spec 2000	Optional	20 cm	800 cm^{-1}
Polytec [4]	FIR–30	Continuous	20 cm	1000 cm^{-1}

We can recall (Eq. 1.41) that resolution can be lost if the drive does not maintain good alignment during the entire scan, and that a poor drive will affect the resolution more at high frequencies than at low frequencies. Therefore, it may not be surprising that most manufacturers demonstrate the maximum specified resolution below 100 cm^{-1} and never, to the knowledge of this author, near the high frequency limit. However, all of these instruments perform very well in the low frequency region of the spectrum.

Not included in this table were any of the rapid-scanning interferometers which will be described in more detail in the next section. Most of these instruments were designed for mid-infrared spectroscopy, but by changing their beamsplitter to a film of Mylar (polyethylene terephthalate), they can also be used successfully for far-infrared spectroscopy. The performance of beamsplitter materials for far-infrared Michelson interferometers will be discussed in more detail later in this chapter.

Most of the interferometers which have been constructed by individual researchers for far-infrared spectroscopy have been rather similar to the commercial instruments listed above, and it certainly worthy of note that both the early interferometers of RIIC and Grubb-Parsons are derived from equipment originally designed by Gebbie's group at the National Physical Laboratory at Teddington, England. Some of the very finest far-infrared measurements have been taken with the NPL "cube" interferometer, which is essentially identical to the Grubb-Parsons Mark II interferometer, the least expensive of all the commercial interferometers. This instrument, with an extended drive to permit measurements at 0.2 cm^{-1} resolution, is shown in Fig. 4.1.

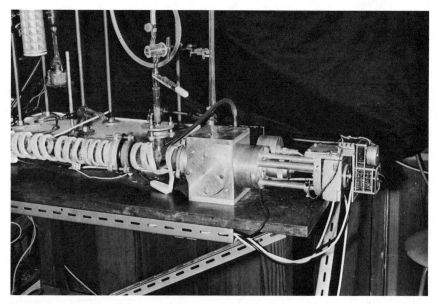

Fig. 4.1.   The NPL "cube" interferometer with an extended (5 cm) drive; the exit beam in this case is passing through a 1 m heated gas cell.

C. MEDIUM RESOLUTION MID-INFRARED
INTERFEROMETERS

The drive mechanism of mid-infrared interferometers has to be somewhat more precise than that of their far-infrared analogs in view of the fact that

$\bar{v}_{max}$ is at least 10 times larger than for the far infrared. Additionally, for absorption spectroscopy, rapid-scanning techniques must be used to keep the dynamic range of the interferogram within the limits of the digitization system, and signal averaging must be carried out if the signal-to-noise ratio is to be increased above that of the single scan case. For exact signal averaging, each interferogram must be sampled at exactly the same retardation interval for every scan. Mid-infrared interferometers must therefore be free of any short- or long-term drifts in alignment, so they are often thermostatted. It is equally important that there should be some kind of "fiduciary marker" to signify the exact retardation at which the *first* data point should be sampled.

All the current rapid-scanning interferometers are direct descendents of the first interferometers designed by Mertz [5]. These early instruments were small and simple, but of rather too low resolution to be really useful for chemical spectroscopy. They derived much of their sensitivity from the large acceptance angle of the interferometer. However, the large divergence angle of the beam together with the low precision drive and low resolution meant that while interferometers of this type were successfully used for several measurements of extended remote sources, they were not used for many laboratory measurements where either the source or the sample was of small size.

The moving mirror of these interferometers is mounted on a spring and displaced by an electromagnetic transducer similar to the voice coil of a loudspeaker. A slowly increasing current is applied to the coil to drive the mirror at a constant velocity. At the end of the scan, the direction of the current is rapidly altered, thereby retracing the mirror to its rest position. The drive current is again reversed and gradually increased so that a second drive cycle begins. The trigger for data collection is a constant frequency clock, so that sampling is strictly linear in time rather than linear in retardation. Nevertheless, in view of the shortness of the scan, this method allowed signal-averaging techniques to be used successfully to increase the signal-to-noise ratio in interferograms from weak sources. A photograph of one of these early interferometers is shown in Fig. 4.2.

The constant frequency time-base which was used for these early rapid-scanning interferometers gave good results primarily because the retardation was so short (<1 mm) that there was not enough time for the interferograms to get out of phase. Had the retardation been much longer, the chances of digitizing the interferogram at corresponding points along the scan (sometimes called *coherent addition*) would have been reduced considerably. For rapid-scanning interferometers designed for higher resolution operation in the mid and near infrared, a method of coherently signal-averaging successive interferograms was developed by Mertz and Curbelo at Block Engineering [6, 7].

Fig. 4.2. The Block Engineering Model 196 interferometer; for several years this was the only rapid-scanning interferometer in common use. (Reproduced from Bulletin 940, Dunn Analytical Instruments, Division of Dunn Associates, Inc., Silver Spring, Md.)

This method involved the use of a second (reference) interferometer, the moving mirror of which was attached to the moving mirror of the main, or signal, interferometer. When the monochromatic light from a laser is passed through this reference interferometer to a detector, a sinusoidal signal is measured. Each zero crossing of this signal can be used to digitize the signal from the main interferometer, since each of these points occurs at equal intervals of retardation. For a 0.6328 μm wavelength helium-neon laser, each zero crossing occurs at retardation intervals of 0.3664 μm. Thus the shortest allowed wavelength in the spectrum would be 0.6328 μm (i.e., $\bar{\nu}_{max} = 15804$ cm^{-1}. This sampling interval could therefore be used if the radiation measured at the detector contained all the frequencies in the infrared.

In general, owing mainly to the properties of beamsplitters, smaller spectral regions are generally covered in any one measurement. For instance, the mid-infrared region of the spectrum is often defined as covering the region between 450 and 3800 cm^{-1}. This bandwidth is more than four times smaller than the region allowed by the sampling interval above, and so a larger sampling interval can be used to reduce the number of data points needed. Generally every second, fourth, eighth, and so forth zero crossing is used to trigger data collection. Thus, if every second zero crossing were used, a bandwidth of $1/2 \times 15804$, or 7902 cm^{-1} can be handled, so that this interval can be used in measurements of the mid-infrared spectrum, as defined above, while for far-infrared spectroscopy, longer sampling intervals can be used.

It should be noted that care must be taken to correctly filter out all high frequency spectral information and noise when long sampling intervals are used so that the computed spectrum does not show additional noise or spectral features due to folding.

The laser fringe-referencing system is not the only additional feature required by signal-averaging types of interferometers, since the sinusoidal signal at the detector of the reference interferometer gives no indication as to where either the zero retardation position of the main interferometer is situated or, more important still, where the first sample point should be recorded. Without this knowledge, the first data point could be recorded at different points on successive scans, and rather than the signal-to-noise ratio of the interferogram being improved on signal averaging, it could actually be degraded.

The technique that is generally used to ensure that the first data point is always sampled at the same retardation involves shining the radiation from a source of white visible light through the reference interferometer, and measuring the interferogram from this source with yet another detector. Since the frequency of the white light is high and its bandwidth is broad, its interferogram will be sharp relative to the signal interferogram. Since the zero retardation position as measured by the white light detector always occurs in the same place, this signal can be used to initiate data collection. By adjusting the position of the fixed mirror of the reference interferometer, the zero retardation point of the white light interferogram can be set at any point along the signal interferogram.

If a single-sided interferogram is being measured, this white light fringe should be set to occur just before the zero retardation point of the signal interferogram. This allows a sufficient number of points to be measured on one side of zero retardation for the small double-sided interferogram required for phase correction but gives most of the data points on the other side of zero retardation so that the desired resolution can be achieved with the minimum scan length and number of data points.

These techniques have been successfully incorporated into a series of interferometers by Block Engineering and its subsidiary Digilab [8]. The first of their interferometers that used this principle had the moving mirrors of the main and reference interferometers back-to-back, using two interferometers similar to the Model 196 with a common drive. Light from the IR source was passed into one interferometer and measured with the appropriate detector. Into the other interferometer was passed monochromatic light from a He-Ne laser (measured with a Ge photodiode) and white light from a small lightbulb (measured with a Si photodiode). The signals from the Ge and Si photodiodes were processed together to give a square-wave signal. The first point on this square-wave corresponded to the first time the white light interferogram reached a certain voltage, and all subsequent variations were generated by the laser signal (Fig. 4.3). The frequency of the square-wave could be set at 1/2, 1, 2, 4 ... of the modulation frequency of the laser.

After this system had been used to verify the validity of the technique, Block Engineering developed other interferometers of higher resolution. The first of these systems was the Model 296 in which the reference interferometer

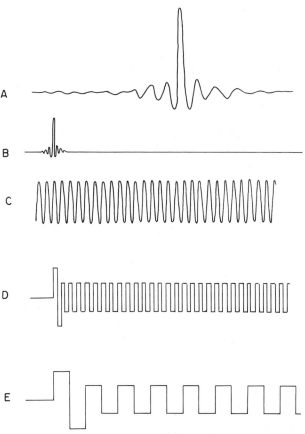

Fig. 4.3. The signals in a typical laser fringe-referenced inter-ferometer: (a) The signal (IR) interferogram; (b) the "white light" interferogram; when this signal reaches a certain threshold voltage, data collection is initiated; (c) the sinusoidal interferogram due to the laser, frequency $\bar{v}_L$: (d) a square wave formed from the laser interferogram; each zero crossing corresponds to one sampling point; (e) a square wave of one quarter the frequency of (d); this sampling frequency is used if shorter spectral ranges than (d) is used for are being studied and can be varied as $\frac{1}{2}\bar{v}_L$, $\frac{1}{4}\bar{v}_L$, and so on.

is a small "cube" which is not back-to-back with the signal interferometer but is still in parallel with it. This arrangement forms the basis of all the Digilab interferometers (Fig. 4.4). The smallest of these laser fringe-referenced interferometers (the Model 197) can achieve a resolution of 4 cm^{-1}, and the largest (the Model 496) will allow a resolution slightly better than 0.1 cm^{-1} to be attained. The drive in the Digilab 496 and 296 interferometers uses an air bearing to ensure a smooth travel, whereas the lower resolution systems use an oil bearing similar to that of the earliest Block interferometers. The mirror velocity is set to give a known frequency output from the laser reference signal. Any variations from this frequency cause the drive voltage to be increased or decreased to rectify the error so that the mirror velocity is essentially constant. This is necessary since major variations in the mirror velocity cause an increase in the noise level of the spectrum and a distortion of the instrument line shape.

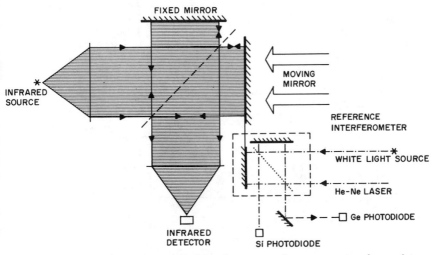

Fig. 4.4. Diagrammatic representation of interferometers where a separate reference interferometer is used to produce the white-light and laser interferograms.

Most of the modern commercial interferometers for mid-infrared spectroscopy, a summary of which is given in Table 4.2, have a similar design to the one shown schematically in Fig. 4.4. However in some interferometers, small regions of the main beamsplitter are coated with a layer of a different material, which eliminates the need for a reference interferometer, as shown in Fig. 4.5. Such a design has the advantage of simplicity, but does not allow the signal interferogram to be easily measured any great distance before

the zero retardation point, since both the signal and the white light interferograms have their maximum value at the same retardation. This design
has been successfully incorporated in the interferometers made by Idealab
Inc. [9].

TABLE 4.2   Commercial Mid-Infrared Interferometers and Fourier
Transform Spectrometers

Manufacturer	Interferometer	Spectrometer	Scan Type	Resolution
Bruker	—	IFS–145	Rapid-scan	0.067 cm^{-1}
Digilab [8]	197	FTS–12	Rapid-scan	4 cm^{-1}
Digilab	—	FTS–10	Rapid-scan	2 cm^{-1}
Digilab	296	FTS–14	Rapid-scan	0.5 cm^{-1}
Digilab	396	FTS–15	Rapid-scan	0.25 cm^{-1}
Digilab	496	FTS–20	Rapid-scan	0.1 cm^{-1}
EOCOM [10]	7001	—	Rapid-scan	0.067 cm^{-1}
Idealab [9]	IF–3	—	Rapid-scan	0.5 cm^{-1}
Idealab	IF–6	—	Rapid-scan	0.1 cm^{-1}
Polytec [4]	—	MIR–30	Rapid-scan	0.7 cm^{-1}
Spectrotherm [11]	—	ST 10	Rapid-scan	4 cm^{-1}
Willey [12]	—	318	Slow-scan	4 cm^{-1}

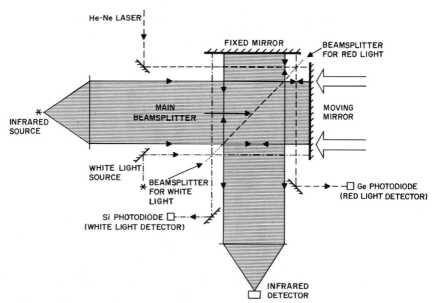

Fig. 4.5. Diagrammatic representation of interferometers which use different areas of a
single beamsplitter for the signal, white-light, and laser interferograms; the area for the signal
beam is at least 90% of the total area.

Recently another commercial interferometer for mid-infrared measurements, which possesses features not found on either the Digilab or Idealab interferometers, has been introduced by EOCOM [10]. The most important feature is the increase in resolution due to the longer stroke of the mirror, providing a maximum retardation of 17.8 cm. The moving element of this interferometer rides on two high-precision air-bearings on stainless-steel rods. The bearings are symmetrical in that the center of resistance is the point at which the drive force is applied, thereby providing for greater drive stability. The feedback control to provide a constant velocity is obtained directly from a tachometer which allows the interferometer to work in an environment with severe vibrations without the need for an accelerometer (*vide infra*, Section IV, D).

The aperture of this interferometer can be made as large as 7.5 cm diameter, as compared to 2.5 cm for the high resolution Digilab and Idealab interferometers, which means that a greater throughput, and hence a greater signal-to-noise ratio, can be obtained at high resolution. Finally interchanging the beamsplitter, which is necessary for the study of different spectral ranges, is simpler on this instrument than on any other mid-infrared interferometer. Beamsplitters are inserted and removed merely by pushing them in or out of the holder, so that only a minor touch-up of the alignment of the fixed mirror is required to optimize the signal for visible light. For longer wavelength operation, realignment can be ignored.

D.   TILT COMPENSATION FOR HIGH RESOLUTION INTERFEROMETRY

To achieve a resolution much higher than 0.1 cm^{-1} at frequencies as high as $10,000 \text{ cm}^{-1}$, extreme care has to be taken over the interferometer drive to avoid the effects described in Chapter 1, Section IX. In theory, there is really no reason why a perfect mirror drive should not be able to be made, but in practice it is rather difficult to move a mirror for long distances (> 5 cm) without accidentally tilting it or shifting it laterally.

If the moving mirror of a Michelson interferometer is tilted, the modulation of the interferogram is reduced. If a corner retroreflector is used as the moving element, the effects of tilt are essentially eliminated (Fig. 4.6). However, accidental lateral displacements produce a shear which has a similar effect on the interferogram. In practice, it is easier to meet the tolerance on lateral displacements than on tilts, and retroflectors have practical advantages over plane mirrors. Steel [13] has suggested a combination of movable retroreflector and stationary mirror that introduces neither shear nor tilt, so that any interferometer using this arrangement would be fully compensated (Fig. 4.7).

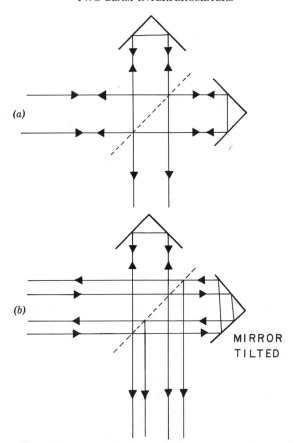

Fig. 4.6. The corner retroreflector: in the upper diagram, (a), both retroreflectors are in good alignment, but in (b) one is tilted. It can be seen that the output beams from each retroreflector are still parallel.

The use of cube corners as retroreflectors has the disadvantage of introducing some polarization effects and requires quite delicate initial alignment. To get around these problems, Connes [14, 15] has used "cats-eyes" as retroreflectors. His systems consist of a concave paraboloidal mirror with a convex sphere being such as to image the beamsplitter back on itself, usually at zero retardation. The principle of the interferometer is shown in Fig. 4.8. In Connes latest spectrometer [16], the interferometer is used in the step-and-integrate mode, so that the complete cats-eye can be stepped from one sample position to the next.

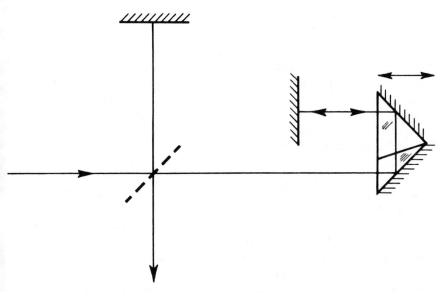

Fig. 4.7. The mirror combination suggested by Steel [13] to compensate for both tilt and shear. (Reproduced from [13] by permission of the author.)

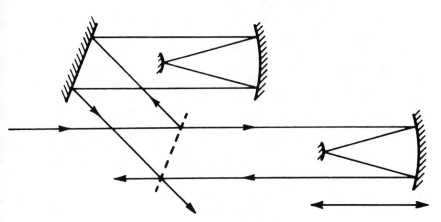

Fig. 4.8. Interferometer using "cats-eye" retroreflectors. (Reproduced from [13] by permission of the author.)

The work of Connes' group at CNRS, Orsay, France represents the peak of achievement to date in the construction of ultrahigh resolution spectrometers. The systems have been described in detail in the literature, and only a brief summary will be given at this point. The cats-eye in the variable path arm of the interferometer is moved on oil bearings, and a maximum retardation of 2 m. can be attained. The operations of displacement and positioning of the moving mirror are continually servo-controlled using a monochromatic source, the path of which occupies the center of the beam under study. Two piezoelectric ceramics support the secondary mirror of the cats-eye and permit very rapid path difference corrections so that up to 50 data points may be measured per second. The secondary mirror of the cats-eye can be vibrated so that the signal is phase modulated, as described in Chapter 1, Section XI. Using this system, a resolution of 0.005 cm^{-1} has been attained using a triangular apodization function. In view of the fact that the Doppler width for spectral lines of gaseous samples is generally of this magnitude, there seems little need for a system of higher resolution.

Sanderson and Scott [17] have described an interferometer designed predominantly for far-infrared spectroscopy which uses the same basic principles as Connes' system. The drive on this interferometer is not servo-controlled since the tolerances are greater than at shorter wavelengths. However provision has been made for the future use of a He-Ne laser for servo-control if the frequency range is extended. This instrument has a maximum retardation of 20 cm.

An interesting interferometer using cats-eye retroreflectors has been described by Schindler [18]. Although most instruments using this type of optical system have been designed for very high resolution spectroscopy, Schindler's interferometer has been designed for medium resolution work ($\geqslant 0.25$ cm^{-1}) under severe vibrational conditions. The optical layout of this system is shown in Fig. 4.9. The use of double-passing serves two purposes. First, it makes the instrument insensitive to lateral displacement of the moving retroreflector optical axis from that of the fixed retroreflector. Second, it gives a change in retardation of four times the motion of the moving element instead of the factor of two found in a classical Michelson interferometer so that a given resolution can be achieved with half the displacement. Since the primary-to-secondary distance is critical, the primary-to-secondary spacer is a fixed silica tube, so that operation over an extended temperature range is permitted.

This instrument is unusual in that it is designed to scan very rapidly, and can advance the moving element every 2 msec so that 48,000 data points can be collected in 1.5 min. For this reason Schindler designed a servo-system with a very short response time; its time constant is less than 100 μsec. Using

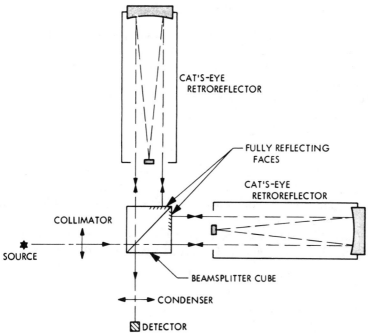

Fig. 4.9. Interferometer designed by Schindler [18] so that a displacement of one cats-eye by a distance x gives a change in retardation of $4x$. (Reproduced from [18] by permission of the Optical Society of America; copyright © 1970.)

this system, the average change in retardation during the 1 msec integration time is less than 2nm. Since this system has been designed to operate under quite severe environmental conditions, it was found necessary to add a reversible fringe counter to keep track of the position in case of temporary loss of servo-lock due to too high an acceleration or vibrational level.

E. CONCLUSION

It is apparent that for any given measurement, a trade-off exists between cost and specifications. Interferometers for medium or low resolution far-infrared spectroscopy can be constructed or bought relatively cheaply. For mid-infrared spectroscopy at a resolution lower than or equal to 0.1 cm^{-1}, interferometers are still available commerically but at an increased cost reflecting their increased sophistication. When high resolution (<0.05 cm^{-1}) is required, no commerical instrumentation is available and the individual researcher must construct his own interferometer.

II. BEAMSPLITTERS FOR MICHELSON INTERFEROMETERS

One of the most important components governing the performance of a Michelson interferometer for infrared spectroscopy is the beamsplitter. In this section the theoretical factors governing the efficiency of beamsplitters are discussed, and the materials that are used in practice are evaluated.

Let us consider a beam of monochromatic radiation of intensity I entering a Michelson interferometer for which the beamsplitter has a reflectance R and a transmittance T. If there is no absorption of radiation:

$$R + T = 1. \tag{4.1}$$

Figure 4.10 shows the intensity of the various beams in the inteferometer if no interference effects occurred. The intensity of the beam that is transmitted to the detector (the d.c. component of the interferogram) is $2RTI$, while that of the beam returning to the source is $(R^2 + T^2)I$. In order that there is no energy loss occurring at the beamsplitter, the sum of the intensities of these two beams must be equal to I. This is easily verified in the following fashion:

$$2RTI + (R^2 + T^2)I = (R^2 + 2RT + I^2)I$$
$$= (R + T)^2 I$$
$$= I$$

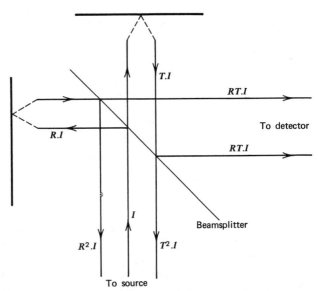

Fig. 4.10. Modified ray diagram of a Michelson interferometer; if light of intensity I enters the interferometer, the output beam going to the detector has an intensity $2RTI$ and the light returning in the direction of the source has an intensity $(R^2 + T^2)I$.

The interferograms of the beams passing to the detector and returning to the source are 180° out-of-phase and their amplitudes about the d.c. level are equal, but of opposite sign, at all moments during the scan. The amplitude of the a.c. portion of the interferogram is determined by the output beam with the lower intensity, which is always the beam which is transmitted to the detector. Figure 4.11 shows how $2RT$ varies with the reflectance of the beamsplitter, and it is seen that $2RT \leqslant 0.5$ for all values of R, so that $2RT \leqslant (R^2 + T^2)$.

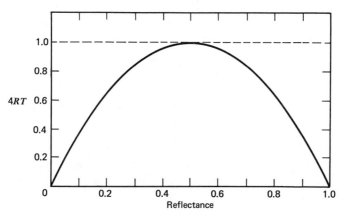

Fig. 4.11.   The variation of $4RT$ with the reflectance, R, assuming that there is no absorption in the film.

Thus for a nonideal beamsplitter, the a.c. component of the interferogram is given by

$$I(\delta) = 2RTI(\bar{v}) \cos 2\pi\bar{v}\delta. \tag{4.2}$$

For an ideal beamsplitter, the interferogram is given by

$$I(\delta) = 0.5\, I(\bar{v}) \cos 2\pi\bar{v}\delta, \tag{4.4}$$

so that the effect of nonideality of the beamsplitter material may be allowed for by multiplying the intensity $I(\bar{v})$ by a factor $\eta(\bar{v})$ which is less than one, that is,

$$I(\delta) = 0.5\, \eta(\bar{v})I(\bar{v}) \cos 2\pi\bar{v}\delta \tag{4.3}$$

$\eta(\bar{v})$ is equal to $4RT$ and is known as the *relative beamsplitter efficiency*. For infrared spectroscopy, $\eta(\bar{v})$ should be as close to unity as possible over as wide a spectral range as possible.

Beamsplitters are generally thin films whose reflectance is determined by the refractive index of the material, n, the thickness of the film, d cm, the angle of incidence of the beam, θ, and the frequency of the radiation, $\bar{v}$ cm^{-1}. Several factors have to be taken into account when calculating the reflectance of the film; apart from the single-surface reflectance, the effect of internal reflections must be considered together with the fact that phase shifts of 180° occur at external reflection surfaces while no phase shifts occur for internal reflection. The ray that is reflected from the front surface of the beamsplitter can interfere with the parallel ray which has undergone internal reflection.

In order to calculate the relative beamsplitter efficiency of any film it is necessary to consider separately the reflectance for radiations whose electric field is polarized parallel to the plane of incidence (p-polarization) and perpendicular to the plane of incidence (s-polarization). The Fresnel equations [19] give the single surface reflectivities for these two polarizations as:

$$R_p = \frac{\sin^2(\theta - \theta')}{\sin^2(\theta + \theta')} \tag{4.4}$$

$$R_s = \frac{\tan^2(\theta - \theta')}{\tan^2(\theta + \theta')} \tag{4.5}$$

where θ' is the angle of refraction inside the film.

Chamberlain et al. [20] have shown that the relative beamsplitter efficiencies for parallel and perpendicular polarized light are given by the following equations:

$$\eta_p = \frac{4R_p T_p^2 E}{(T_p^2 + R_p E)^2} \tag{4.6}$$

$$\eta_s = \frac{4R_s T_s^2 E}{(T_s^2 + R_s E)^2} \tag{4.7}$$

where $T_p = 1 - R_p$ (4.8)

$$T_s = 1 - R_s \tag{4.9}$$

and $E = 4\sin^2(2\pi n d \bar{v} \cos \theta')$

$$\equiv 4\sin^2 \varepsilon \tag{4.10}$$

where $\varepsilon = 2\pi n d \bar{v} \cos \theta'$.   (4.11)

The relative beamsplitter efficiency of a film for unpolarized incident radiation is merely given by the average of the values for parallel and perpendicular polarized radiation:

$$\eta = \tfrac{1}{2}(\eta_p + \eta_s). \tag{4.12}$$

From Eqs. 4.6 and 4.7 it is seen that η will take a zero value when E is zero, that is, when

$$\varepsilon = m\pi$$

where $m = 0, 1, 2\ldots$, and so forth, and a maximum value when $m = \pi/2$, $3\pi/2$, $5\pi/2$, and so forth. The efficiency of any beamsplitter is zero at 0 cm^{-1} and returns to zero for the first time when $\varepsilon = \pi$. Thus the frequency at which this minimum value of $\eta(\bar{v})$ is seen, $2\bar{v}_0$ cm^{-1}, is given by

$$2\bar{v}_0 = (2nd \cos \theta')^{-1} \tag{4.13}$$

Between 0 and $2\bar{v}_0$ cm^{-1}, the variation of $\eta(v)$ is symmetrical about v_0, although the actual shape of the curve is dependent on the values of n and θ, since it is these two parameters that determine the amount of internal reflection in the film.

For materials of low refractive index with a low angle of incidence, the proportion of the incident radiation that is internally reflected is small, and $\eta(\bar{v})$ varies with frequency in the manner shown in Fig. 4.12. The most commonly used beamsplitter material for far-infrared Fourier transform spectroscopy is polyethylene terephthalate (Mylar in the U.S., Melinex in Europe)

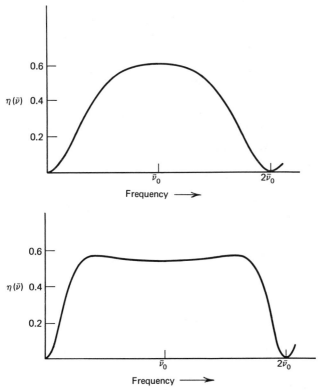

Fig. 4.12. The variation of $\eta(\bar{v})$ with frequency, $\bar{v}$, for a Mylar film, (a) if $\theta = 45°$ and (b) if $\theta = 70°$.

for which the refractive index is 1.69 in the far infrared [20]; Fig. 4.12a specifically refers to a Mylar film used at an angle of incidence of 45°. It can be seen from the figure that for this angle of incidence, $\eta(\bar{\nu})$ for Mylar never exceeds 0.6, due primarily to the low single-surface reflectances at 45° incidence for parallel ($R_p = 0.02$) and perpendicular ($R_s = 0.14$) polarized radiation.

Experimentally it is found that for a film to be useful as a beamsplitter for infrared spectroscopy, $\eta(\bar{\nu})$ should be at least 0.3 and preferably greater than 0.5. Therefore, if the former criterion is applied, the spectral range for a Mylar film àt 45 incidence is between $0.35\bar{\nu}_0$ and $1.65\bar{\nu}_0$. In practice, it is desirable to attain a wider spectral range than this if possible.

The effective spectral range may be increased either by increasing n or θ. If the angle of incidence of a beam of radiation on a Mylar film is increased to 70°, the shape of the efficiency curve changes such that, although the *maximum* efficiency is slightly *lower* than for 45° incidence, the efficiency remains greater than 0.5 over a much wider frequency range (Fig. 4.12b). The value of $\eta(\bar{\nu})$ is greater than 0.5 for $0.25\bar{\nu}_0 < \bar{\nu} < 1.75\bar{\nu}_0$ while it is greater than 0.3 for $0.15\bar{\nu}_0 < \bar{\nu} < 1.85\bar{\nu}_0$. In spite of the reduced maximum efficiency, Mylar is obviously preferable for spectroscopy over wide spectral ranges when it is used with high angles of incidence.

On the other hand, if a high relative efficiency is desired over a small spectral range it becomes preferable to reduce the angle of incidence below 45°. The manner in which the beamsplitter efficiency varies with ε for different incident angles is shown in Fig. 4.13. The minimum value of θ is determined by the Brewster angle below which there would be total internal reflection.

The results for Mylar represent the typical case for a low refractive index material. If the angle of incidence is held constant at 45° and the refractive index is increased, the shape of the beamsplitter efficiency curve becomes similar to that of Fig. 4.12b, and in addition the maximum value of $\eta(\nu)$ becomes closer to unity. The film therefore becomes better suited for infrared spectroscopy, where a high efficiency over a wide spectral range is required.

Since very thin films are required for mid-infrared spectroscopy and also since the high index materials tend to be very brittle, they have to be supported on a transmitting substrate. These substrates should have several important properties:

1. They should have a low refractive index to prevent a large reflection loss from the front surface of the plate.

2. They should have a high transmittance over the entire spectral range of interest and show no strong absorption bands.

3. They must be able to maintain a flatness equal to a quarter of the shortest wavelength being measured.

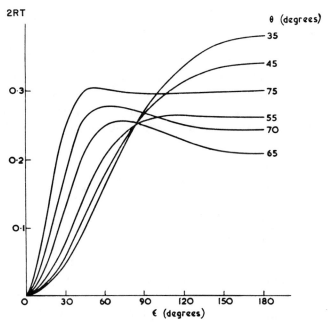

Fig. 4.13.   The variation of 2RT with ε for different angles of incidence for a Mylar film. (Reproduced from [20] by permission of the author and Pergamon Press; copyright © 1966.)

4. They should not be easily scratched or be susceptible to attack by any component of the atmosphere (especially water vapor).

To prevent "chirping" occurring (see Chapter 1, Section VI) a compensator plate of exactly the same thickness as the substrate must be placed on the other side of the film.

Germanium ($n = 4.0$) and silicon ($n = 3.4$) films are generally used for mid-infrared spectroscopy and the plate on which the film is deposited depends on the spectral range to be studied. CsI and CsBr will allow the range to be extended into the far infrared but both materials are soft and do not maintain their flatness for high-frequency measurements. NaCl is a good material as far as hardness is concerned but does not have a high transmittance below 650 cm^{-1}. The best compromise between these two extremes for conventional mid-infrared spectroscopy is KBr. For near-infrared spectroscopy a ferric oxide film ($n = 3.0$) is usually deposited on a calcium fluoride or quartz flat.

Calculations of the beamsplitter efficiency of deposited films are somewhat more complex than for unsupported films since the refractive index of the substrate must be taken into account. Sakai [21] has calculated the efficiency of a typical film with the angle of incidence equal to 45°, the refractive index of the film equal to 3.6, and that of the substrate equal to 1.4 (Fig. 4.14). It can be seen that the relative efficiency is greater than 0.8 for frequencies between $0.3\bar{v}_0$ and $1.7\bar{v}_0$, but outside this range $\eta(\bar{v})$ drops off rapidly. For mid-infrared spectroscopy ($\bar{v}_0 = 2000$ cm^{-1}) using this beamsplitter, good performance may be expected between 600 and 3400 cm^{-1} but the performance will be decreased at frequencies above 3400 cm^{-1} and below 600 cm^{-1}.

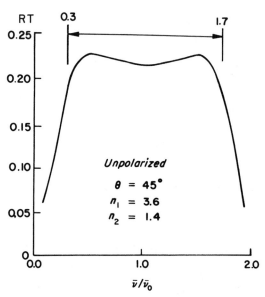

Fig. 4.14. The variation of RT with frequency for a silicon film on a sodium chloride substrate with a NaCl compensator plate of the same thickness; $\theta = 45°$. (Reproduced from [21] by permission of the author.)

The Fresnel equations show that the relative beamsplitter efficiency of a given film is different for parallel and perpendicular polarized radiation. The reflectance for p-polarized light is always less than that for s-polarization, but this fact does not imply that $\eta(\bar{v})$ is always lower for parallel polarized radiation if the beamsplitter material has a high refractive index. For instance, Bell [22] has given the overall reflectance of germanium at $\bar{v}_0$ as 0.36 for p-polarized radiation and 0.62 for s-polarization. In this case the relative efficiencies $[= 4R(1 - R)]$ at $\bar{v}_0$ for each polarization are given by $\eta_p(\bar{v}_0) = 0.92$ and $\eta_s(\bar{v}_0) = 0.94$. Thus, even though the reflectance values are quite different, the fact that they differ from the ideal value of 0.5 by approximately the same extent causes the relative efficiencies for each polarization to be approximately equal.

Figure 4.15 shows the variation of the overall maximum reflectance for materials of different refractive index. Since an ideal beamsplitter has a reflectance of 0.5, it is apparent from this diagram that the optimum beamsplitter material for an unpolarized beam should have a refractive index of about 3.6, whereas if only *p*-polarized radiation is being studied the ideal refractive index is about 2.8.

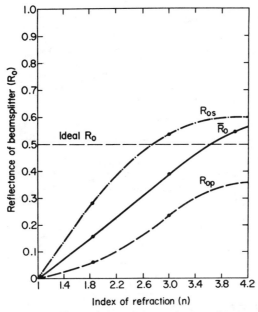

Fig. 4.15.   The variation of the maximum reflectance, R_0, of a beamsplitter film with refractive index; the optimum value of $\eta(\bar{\nu})$ is found when $R = 0.5$, so that an ideal beamsplitter cannot be found for p-polarized radiation. (Reproduced from [22], p. 119, by permission of the author and Academic Press; copyright ©️ 1972.)

Tescher [23] has described the results of a theoretical investigation into the effect of polarization on $\eta(\bar{\nu})$ for dielectric films on transmitting substrates, with compensator plates of the same material. His results for a 0.4 μm film of germanium between NaCl plates are shown in Fig. 4.16*a* and for a 0.17 μm film of silicon between CaF_2 plates in Fig. 4.16*b*. These systems show good

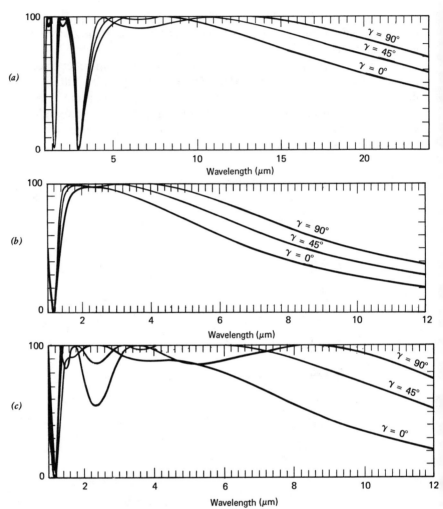

Fig. 4.16. Theoretical calculations of beamsplitter efficiency for various combinations of materials at different incident angles: (a) a 0.4 μm film of germanium between NaCl plates; (b) a 0.17 μm film of silicon between CaF_2 plates; (c) a 0.17 μm film of silicon between CaF_2 plates with a 2.5-μm air gap between the film and the compensator plate; note the improved long wavelength performance. (Reproduced from [23] by permission of the author.)

performance over fairly limited spectral ranges but the performance falls off at long wavelength.

By adding an additional thin film layer a substantial improvement in the long wavelength performance can be achieved. If the CaF_2 compensator plate used for Fig. 4.16b is separated from the silicon film by a 2.5 μm air gap, it is seen that $\eta(\bar{\nu})$ is increased at long wavelength (Fig. 4.16c). Tescher also demonstrated that the width of the air gap can be varied to optimize the performance in different spectral regions.

In practice, it is difficult to control the thickness of an air gap, and Tescher has also described a system where 13 layers of different materials (Ge, TlBr, KRS–5) can be deposited to produce a beamsplitter with $\eta(\bar{\nu}) > 0.5$ from 7000 to 300 cm^{-1}. Although these beamsplitters would obviously be difficult to manufacture since the thickness of each layer must be carefully controlled, for any application where a large bandwidth is more important than good response over a small frequency range, systems such as this seem to hold the highest potential.

At the present time, stretched polymer films are used most commonly for far-infrared Fourier transform spectroscopy while thin films of a single material on transmitting substrates are generally used for mid-infrared measurements. Several workers have studied the possibility of extending the use of unsupported polymer films as beamsplitters for mid-infrared spectroscopy. From the earlier discussion of these materials, it is apparent that their principal advantage rests in their cheapness and ease of handling rather than high efficiency over wide spectral ranges. Thorpe et al. [24] have described attempts to extend the range of a Beckman-RIIC FS–720 inter-ferometer to 1200 cm^{-1}. These workers used not only Mylar beamsplitters (which have several strong absorption bands below 1200 cm^{-1}) but also polyethylene and polypropylene which have fewer absorption bands, but whose refractive index (approximately 1.5) is even lower than that of Mylar, which makes them even less suitable for use as beamsplitters.

In order for an unsupported plastic film to be effective as a beamsplitter, a thickness of at least 2 μm appears to be necessary for reasons of availability and mechanical strength. This thickness places an upper limit of approxi-mately 1200 cm^{-1} for all polymer films working in the first order ($\varepsilon < \pi$). Since all polymers have strong absorption bands in the region between 200 and 1200 cm^{-1}, no single material can be used to provide coverage across this entire frequency range. In an attempt to increase the efficiency over a *short* frequency range, Thorpe's group used thicker films whose relative efficiency was at a maximum in the frequency range being studied. They obtained their best results above 700 cm^{-1} with 12.5 and 25 μm thick films of polyethylene and polypropylene around frequencies for which ε was equal to $3\pi/2$, $5\pi/2$, ... etc. Even with these films, the maximum efficiency is considerably less than 0.5, due not only to the low refractive index of these materials but also to the amount of absorption and scattering in the film.

Spectra measured in this manner do not compare at all favorably with spectra measured using silicon or germanium beamsplitters. A second major disadvantage of thick polymer films used in higher orders is the very short spectral range that is able to be covered. In view of the low efficiency and short frequency range, all advantages of an interferometer over a grating spectrometer with the same optical throughput are lost.

One type of material other than dielectric films may become useful as a beamsplitter for far-infrared spectroscopy. Wire grids have been shown by Vogel and Genzel [25] to possess the properties of reflectance and transmittance for far-infrared radiation which give quite high relative beamsplitter efficiencies. The efficiency of a wire grid at any wavelength, λ, depends on the grating constant, d, of the mesh but is not strongly dependent on the width of the wires. A plot of $\eta(\bar{v})$ against λ/d is shown in Fig. 4.17, which demonstrates that high efficiencies can be attained over reasonably long spectral

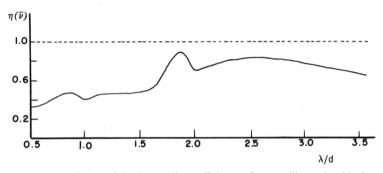

Fig. 4.17. Variation of the beamsplitter efficiency of a metallic mesh with the parameter λ/d, where λ is the wavelength of the radiation and d is the grating constant of the mesh.

ranges. Although this type of beamsplitter shows no advantages over high refractive index dielectric films, it appears to be superior to Mylar for far-infrared spectroscopy and it seems surprising that wire grids have not been more extensively used for this spectral region.

III. LAMELLAR GRATING INTERFEROMETERS

Although it is probably true at this time to say that more Michelson interferometers have been used for measurements below 200 cm^{-1} than for any other spectral region, this type of instrument has a few disadvantages for very far-infrared spectroscopy. The main disadvantage of the Michelson interferometer as it is conventionally used for far-infrared spectroscopy is the low efficiency and limited spectral range given by the Mylar beamsplitter. On the other hand the type of interferometer known as a *lamellar grating interferometer* can have an efficiency close to 100% and can operate over a wide spectral range for far-infrared spectroscopy [26]. Although the optical and mechanical properties of lamellar grating interferometers preclude their use at frequencies much above 150 cm⁻¹, they have been successfully used from 150 cm⁻¹ to as low as 1.5 cm⁻¹. For all measurements below 10 cm⁻¹, the lamellar grating interferometer is almost certainly the instrument of choice.

Unlike the Michelson interferometer, in which the amplitude of the radiation is divided at the beamsplitter, the lamellar grating interferometer uses the principle of *wavefront division*. It consists of two sets of parallel interleaved mirrors, one set of which can move in a direction perpendicular to the plane of the front faces, while the other set is fixed. Thus, if a beam of collimated radiation is incident onto the grating (Fig. 4.18), half the beam will be reflected from the front facets while the other half will be reflected from the back facets. By moving one set of mirrors, the path difference between the two beams can be varied, so that an interferogram is generated. It can be seen from Fig. 4.18 that essentially all the radiation from the source reaches the detector, so that the efficiency of the lamellar grating interferometer can approach 100%.

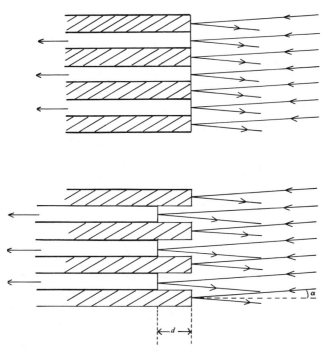

Fig. 4.18. Principle of the lamellar grating interferometer; radiation reflected off the movable facets travels a distance of $2d/\cos \alpha$ further than the radiation reflected off the stationary (shaded) facets.

In practice the efficiency can only approach 100% through a limited frequency range, and it falls off at both high and low frequencies [26]. Figure 4.19 shows the theoretical efficiency of a lamellar grating interferometer compared with that of a Michelson interferometer using a Mylar beamsplitter. The low frequency cut-off is caused by the *cavity effect* which begins to decrease the modulation of the wave for which the electric vector is parallel to the sides of the cavity. The frequency $\bar{v}_L$ is equal to $(0.3a)^{-1}$ where a is the grating constant. The reduction in efficiency at high frequency is caused by waves diffracted at the grating (which can be considered as a series of long rectangular slits) being cancelled at the exit aperture. The frequency $\bar{v}_C$ is equal to F/aS, where F is the focal length of the collimator and S is the diameter of the exit aperture.

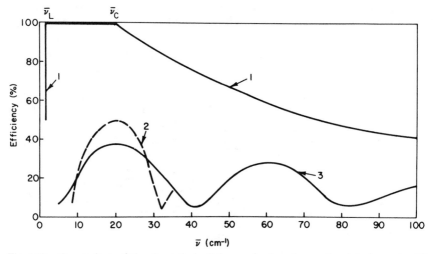

Fig. 4.19. Comparison of the theoretical efficiency of a lamellar grating interferometer (1) and a Michelson interferometer with (2) a metal mesh beamsplitter and (3) a Mylar beamsplitter. (Reproduced from [26] by permission of the author and the Optical Society of America; copyright ©️ 1964.)

Figure 4.20 shows how the actual energy measured with a lamellar grating interferometer compares with the energy of the same source measured with a Michelson interferometer equipped with a 100 μm and 50 μm thick Mylar beamsplitter [1], using the same detector. Below 10 cm^{-1} the energy in the single-beam spectrum measured with the lamellar grating interferometer is decidedly superior to that measured even with the thickest Mylar film. However for use above 100 cm^{-1} the lamellar grating interferometer requires an exceptionally high quality drive and in practice it becomes preferable to use a Michelson interferometer.

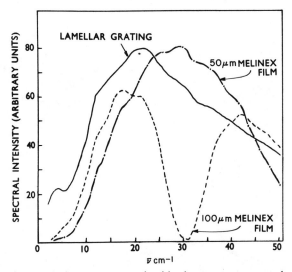

Fig. 4.20. Spectra measured with the same source and detector taken using a lamellar grating interferometer and a Michelson interferometer with 100 and 50 μm thick Mylar beamsplitters. Note the better low-frequency performance of the lamellar grating interferometer.

Lamellar grating interferometers are always used with the beam slightly off-axis. If this were not the case most of the energy would be lost, since it would have to return to the source. Strong [27] originally used a system with an Ebert mirror for his first measurements (Fig. 4.21a). In the Beckman-RIIC Model LR–100 [1] (the only commercial lamellar grating instrument), a somewhat more on-axis system is used, and this is shown in Fig. 4.21b. Bell [22] has suggested the use of two light-pipes to pass the radiation into and out of the interferometer. The beam is collimated by the large spherical mirror, A, and reflected to the lamellar grating. The beam returns to A after a phase difference has been generated between the beams passing to the front and back facets, and is refocused on the exit light-pipe which is situated next to the entrance light-pipe. This arrangement is shown diagramatically in Fig. 4.21c. The designs shown in Figs. 4.21b and c keep the beam close to the optical axis of the collimator, thereby reducing abberations, astigmatism, and "shadowing" of the back facets by the front ones (Fig. 4.22). The magnitude of all three effects is greater at large displacements of the facets, and each serves to reduce the modulation of the beam. Thus, each effect optically apodizes the interferogram so that the instrumental line shape from a lamellar grating interferometer at a given retardation is broader than for the same

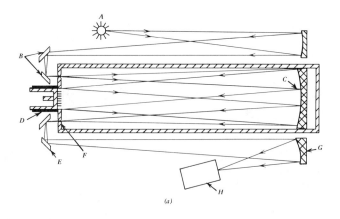

(a)

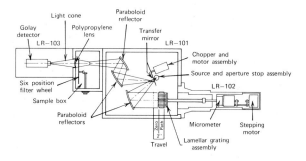

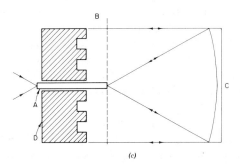

(c)

Fig. 4.21. Various designs for lamellar grating interferometers: (a) the interferometer designed by Strong; the mirror C is an Ebert mirror and the exit beam is focused onto the detector H by the off-axis ellipsoid G. (From *Concepts of Classical Optics* by John Strong, W. H. Freeman and Company. Copyright © 1958.) (b) the Beckman–RIIC Model LR–100 lamellar grating interferometer; (c) the design suggested by Bell which has two light-pipes (A) to bring the beam into and out of the interferometer; one light-pipe is placed over the other and their ends are in the focal plane of the collimator mirror (C). This arrangement keeps the beam close to the optical axis of the collimator thereby reducing aberrations and shadowing. (Reproduced from [22], p. 206, by permission of the author and Academic Press. Copyright © 1972.)

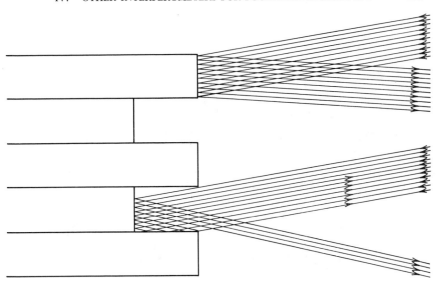

Fig. 4.22.   The effect of shadowing; all the light hitting the front facets is reflected towards the detector, whereas some of the light from the rear facets is reflected back in the direction of the source, thereby reducing the efficiency of the interferometer.

retardation using a Michelson interferometer. Since the magnitude of these effects increases as the solid angle of the beam increases, the possible through-put advantage of a lamellar grating interferometer is not as great as that of a Michelson interferometer. However, in the far infrared very few interferometers can be used with a throughput corresponding to the solid angle allowed in theory because of the small size of the source or detector, so that this effect is generally not very important. Thus lamellar grating interferometers show their principal benefits for relatively low resolution spectroscopy in the very far infrared.

To increase the throughput of a lamellar grating interferometer, Hansen and Strong [28] designed a system using spherical mirrors (Fig. 4.23). In this instrument a very highly convergent beam of light may be passed to the grating, so that a higher throughput can be achieved than for either a conventional Michelson interferometer or a plane lamellar grating interferometer.

IV.   OTHER INTERFEROMETERS FOR FOURIER SPECTROSCOPY

A.   INTRODUCTION

There are several modified versions of the basic Michelson interferometer which have been designed for a variety of special purposes. The more important of these will be described in the following sections. For example, the field-widened interferometer has been designed to accept a greater solid angle

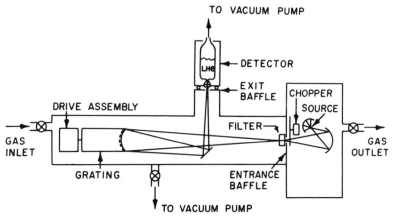

Fig. 4.23. Lamellar grating interferometer with spherical mirrors; this configuration eliminates several of the focussing mirrors found in most lamellar grating interferometers. (Reproduced from [28] by permission of the author.)

than is allowed for the conventional Michelson interferometer, and so can be used for measuring the spectra of extended sources. An interferometer with spherical mirrors rather similar to the lamellar grating interferometer described at the end of the last section is also described. Cryogenic interferometers permit measurements to be made at a previously unattainable sensitivity. Specially modified or designed interferometers for operation in mobile environments such as aircraft and space vehicles are also described.

B. FIELD-WIDENED INTERFEROMETERS

Although most spectroscopic measurements are made with small sources or samples, there are certain types of emission measurements where the source can be quite large. In particular, for measurements of twilight and night airglow the source is very extensive and extremely weak. Thus the signal-to-noise ratio in the interferogram could be improved if the solid angle of the radiation passing through the interferometer could be increased without sacrificing resolution.

The reason why the acceptance angle for a given resolution has to be kept below a certain limit was discussed in Chapter 1, Section VII. If some means of making the path difference invariant with the entrance angle were found, then a greater throughput from an extended source could be achieved, and hence a greater signal-to-noise ratio attained.

A technique suggested by Bouchareine and Connes [29] accomplishes such field compensation to the first order by using prisms in place of the usual mirrors of the Michelson interferometer as shown in Fig. 4.24. The back sides of the prisms are silvered, and the optical system is aligned as in a typical interferometer. Instead of obtaining differential pathlengths by moving one mirror parallel with the rays of light that impinge on it, in the field-widening technique a prism is moved parallel with its "apparent mirror position," inserting more material in one beam to effect an increase in retardation. "Apparent mirror position" refers to the apparent position of the coated back surface of the prism, looking through the optical material. Thus the direction of the drive is perpendicular to the light rays. The fact that the retardation is independent of the incident angle will not be proved here, but it can be shown to follow from Fermat's principle. To achieve the field-widening increase in optical throughput, large prisms must be constructed and displaced during

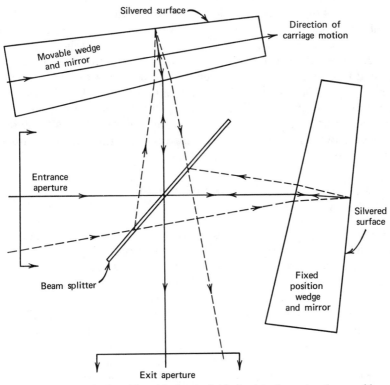

Fig. 4.24.   The method used for widening the field of an interferometer; the movable rear-silvered prism is translated in a direction perpendicular to that of the movable mirror in a Michelson interferometer, so that the retardation is increased by increasing the distance travelled through the prism. (Reproduced from [30] by permission of the author.)

operation, so that the wavefront distortions are less than $\lambda/10$ across the full travel of the prism.

Despain et al. [30] have described an interferometer using this principle for operation in the near infrared and visible regions of the spectrum. They have used a prism with a wedge angle of 8° constructed from quartz because of its low dispersion and expansion coefficient. With this particular system, they have shown that the retardation, δ, is related to the distance, x, over which the wedge is moved by the equation:

$$\delta = 0.21x.$$

Thus to achieve a resolution of 1 cm^{-1} a drive length of at least 4.7 cm is needed. To maintain accurate positioning of the prism over the full length of this scan presented a real problem in stability. The base of the interferometer had to be constructed out of granite, and the optical carrier was made out of invar stainless steel, whose very low expansion coefficient is close to that of granite and quartz.

This instrument can be used with either a photomultiplier or a photo-conductive detector. When a photomultiplier is used the ratio of the times required to measure spectra between 0.4 and 0.7 μm at the same signal-to-noise ratio using a standard Michelson interferometer and the field-widened interferometer is about 19:1. If a photoconductive detector is used to measure spectra at 2.5 μm with the same signal-to-noise ratio using standard Michelson and field-widened interferometers, the ratio of the measurement times is increased to about 360:1.

C. MICHELSON INTERFEROMETER WITH SPHERICAL MIRRORS

Bottema and Bolle [31] have described an unusual interferometer designed for far-infrared spectroscopy of weak small sources. Normally there would be no advantage in using anything other than a standard lamellar grating or Michelson interferometer for this purpose, but since their interferometer had to be mounted in a rocket, strong consideration had to be given to the size and weight of the system. To reduce the number of collimating mirrors before the interferometer and mirrors required to focus the exit beam onto the detector, a Michelson-type interferometer was built with spherical rather than plane mirrors (Fig. 4.25). The focus of the beam passing from the moving mirror to the detector changes with the position of the mirror, but Bottema and Bolle found that the usable path difference was quite long, being equal to twice the focal tolerance in the image, formed by either mirror, of its center of curvature.

The system described by Bottema and Bolle was designed to accept an $f/6$ beam from a telescope. The radius of curvature of the mirrors was 50 mm

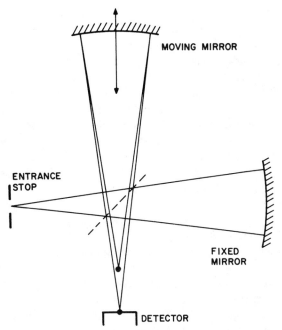

Fig. 4.25. This Michelson interferometer with spherical mirrors was designed to eliminate most of the mirrors normally used to collimate the beam and focus onto the detector. It has been successfully used for far-infrared spectroscopy but would be of less use for mid- and near-infrared measurements. (Reproduced from [31] by permission of the author.)

and the maximum retardation was ± 3 mm. The resolution of this system was compared with that of a plane Michelson interferometer by using a mercury lamp source and Golay detector with an equal pathlength of water vapor between them, and essentially identical interferograms were produced with no loss of modulation at high retardation.

Such a system does not appear to be of very wide application in view of the resolution limitation, but achieves the objectives for which it was designed admirably.

D. MOBILE INTERFEROMETERS FOR REMOTE SPECTROSCOPY

Many of the earliest interferometers used for FTS were designed for operation away from the laboratory, and now interferometers have been used for a wide variety of remote measurements. They have been located outside buildings, in trucks, on tripods and in aircraft, balloons, rockets, and space-craft. Most of the instruments designed for remote sensing have telescopes on the front end of the interferometer to restrict the field-of-view so that the spectra of small remote objects can be measured.

In order for any instrument to be useful for remote measurements, it must be portable. Thus, many of the first measurements were carried out with the small early rapid-scanning interferometers made by Block Engineering. The type of remote sources monitored were smokestack emissions [32] (where the instrument is off-horizontal but stationary), rocket exhausts [33] (where the source is moving rapidly and has to be tracked), and even nuclear deto-nations [34] (where the source is transient and the interferometer is airborne). The same type of interferometer was also sent up in balloons for monitoring terrestrial radiance [35]. In view of the simplicity of this early interferometer, little modification was needed to the basic system described in Section I for any of these measurements.

As an example of the power of interferometry for remote sensing, spectral measurements of a high altitude nuclear detonation made from an aircraft carrying both a monochromator and an interferometer may be compared [34]. The monochromator was a large $f/4$ Ebert-Fastie spectrometer with a 50-cm diameter mirror and a 20×30 cm grating, which was barely able to be loaded through the cargo door of the aircraft; the interferometer was an early Block Engineering instrument with 1-cm diameter mirrors. Although no spectral data were obtained with the monochromator, good emission spectra of the fireball were measured using the interferometer. Vibration isolation for these measurements was carried out by hanging the inter-ferometer in rubber trampoline mounts. However, as resolution require-ments increased, more sophisticated methods for vibration isolation became necessary.

High resolution laser fringe-referenced interferometers generally have a servo-mechanism to control the drive voltage so that the mirror velocity remains constant. If the mirror slows down so that the frequency of the signal from the laser interferometer decreases, the drive voltage is increased so that the velocity is returned to its correct value. This type of system works adequately to compensate for the small variations of velocity that can be found in a laboratory environment. However, under conditions of high acceleration or vibration which can be encountered when the system is held in a field environment, it is possible that the mirror position can change so rapidly that one or more fringes can pass across the detector before the servo-mechanism has a chance to rectify the situation. For this type of measurement an accelerometer can be fixed to the moving mirror to sense any change of acceleration as it takes place. The drive voltage can be ad-justed more quickly in this fashion than could be achieved using the servo-mechanism, and measurements are able to be made when the interferometer is subjected to moderate vibrations. Off-horizontal operation of the interferometer also causes a force tending to accelerate or retard the moving mirror due to the effect of gravity on the mirror, and an accelerometer is used to correct this effect as well.

Many of the remote measurements that are currently being performed use commercially manufactured instrumentation (Block Engineering, Digilab and Idealab) modified to function under the conditions of the test. However, several other groups have developed their own equipment for the particular measurements in which they are interested. The special features of some of these systems will now be briefly discussed.

One of the more difficult problems associated with the use of interferometers in a field environment is the ability to maintain the optical alignment over an extended period of time. Ashley and Tescher [36] have described an interferometer featuring a piezoelectric mirror alignment scheme coupled with three laser reference beams and phase sensitive comparators. The laser is split into three beams and arranged to form an equilateral triangle centered about the main interferometer channel. The output of each beam, modulated by the scanning interferometer, is detected by a photodiode. A phase comparison is made between two of the channels with the third as reference. The phase comparators provide an output voltage that is proportional to the phase difference over a $\pm 360°$ angular error. The resulting voltage developed by any phase error, which corresponds to a misalignment of the optical cube, is amplified and applied to the appropriate piezoelectric transducer. The same group is developing a piezoelectric mirror drive system designed to eliminate all effects introduced by vibrations encountered in a field environment.

One of the most successful experiments to be carried out using a Michelson interferometer in space involved the Infrared Interferometer Spectrometer (IRIS) flown on the Nimbus III satellite [37]. This instrument was designed to have a radiometric precision of 1%, which requires not only good mirror alignment but also an extremely precise mirror drive. The technique used to achieve this precision involves comparing the phase of the monochromatic reference signal to that of a clock frequency derived in the spacecraft. A signal proportional to the phase error is fed back to the drive transducer, so that the drive is under closed-loop operation.

This brief summary has been intended to give an idea of problems involved in the design and operation of interferometers for use in mobile environments. This account has only covered interferometers working at temperatures at or only a little below room temperature. For increased sensitivity, especially for space spectroscopy, the interferometer has to be cooled, and two cryogenic interferometers are described in the next section.

E. CRYOGENIC INTERFEROMETERS

When emission spectra are to be measured from cool sources, radiation emitted from the interferometer itself may swamp the radiation of interest. Therefore, it is sometimes necessary to cool the interferometer to a temperature where its emitted radiation is negligible in comparison to that from the source being studied.

Besides the effect of radiation emitted from the interferometer, there is another important reason [38] to cool down the complete system when certain types of detectors are used. Extrinsic photoconductive detectors such as mercury- and copper-doped germanium as well as mercury-cadmium telluride detectors when cooled to low enough temperatures become almost wholly photon flux noise limited. Thus one would expect that in this case a corresponding situation to that of photomultiplier could exist, namely that the multiplex advantage of the interferometer could be cancelled out by the increased photon flux noise. However, unlike photomultipliers these longer wavelength photodetectors do not possess a reasonably noisefree interval gain mechanism. Therefore, under low background conditions, a situation is approached where further reduction of the background photon flux does not correspondingly improve the detectivity of the detector, so that one becomes limited by other noise sources such as preamplifier noise, vibrations, microphonics and boil-off noise. In this case the multiplex advantage of $\sqrt{N}$ (where N is the number of spectral resolution elements) over dispersive spectrometers is found to apply.

However, one other factor must be borne in mind. The phenomenon of dielectric relaxation causes the detector time constant to become very long for low background flux conditions. Since this dielectric relaxation time constant is inversely proportional to the photon flux upon the detector, this effect produces a situation that when one operates the detector at frequencies in excess of the cut-off frequency of the detector, the system gain (and therefore the signal-to-noise ratio) is directly proportional to the photon flux upon the detector. Therefore one finds that if one is system noise limited, the interferometer can show a gain in signal-to-noise ratio over dispersive spectrometers of as much as $N\sqrt{N}$.

To reach the situation of being system noise-limited, Engel et al. [38] have described how a standard Block Engineering Model 296 interferometer was modified so that it could be cooled down to 77 K. Since the alignment of the mirrors in the main and reference interferometers will change as the temperature is reduced from room temperature, the most important modifications of the system provided for remote alignment of the main interferometer and a reference system free from realignment problems.

An independent reference interferometer has been built into the instrument behind the moving mirror. The use of retroreflectors in this system eliminates the need for realignment after initial adjustment at room temperature since the entire unit including the white light and the laser operates at 77 K.

Adjustment of the fixed mirror in the main interferometer is provided for by a combination of mechanical and piezoelectric elements. The system is aligned initially at room temperature and alignment is maintained through the cooling cycle by varying the voltage applied to the piezoelectric element located under the fixed mirror.

The optical elements are all held at 77 K except for an optical filter and the detector which are both held at 4 K. No modifications were made to the gas bearing for the moving mirror, which was found to perform well at the temperature of liquid nitrogen. A diagram of the system is shown in Fig. 4.26. No performance figures are yet available for this instrument.

Another description of how a commercial interferometer has been modified to allow it to be cooled with liquid nitrogen has recently been given by Moehlmann et al. [39]. Like Engel and his co-workers, these authors also had the goal of measuring extremely low intensity emission spectra for which the statistical fluctuations in the background radiation hitting the

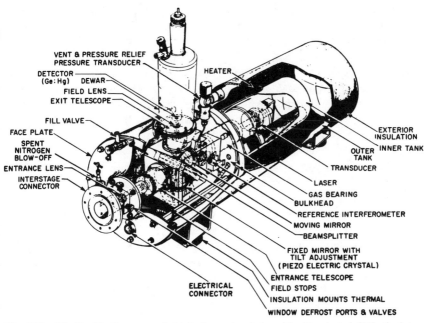

Fig. 4.26.   The liquid nitrogen cooled interferometer designed by Engel et al. [38]; the interferometer for this unit is a standard Digilab Model 296, modified for remote tilt adjustment using a piezoelectric crystal. (Reproduced from [38] by permission of the author.)

detector from room temperature surroundings would have caused a signal-to-noise ratio of 1/50 to have been measured in their experiment. By cooling the interferometer from 300 K to 77 K, the background radiation was reduced by a factor of 10^6. These workers mounted a Digilab Model 296 interferometer in a housing jacketed by liquid nitrogen which cools the entire assembly to 80 K by conduction through the atmosphere of gaseous helium inside the chamber. The optics of the Model 296 interferometer were modified for cryogenic work by remounting the aluminized quartz mirrors on Invar steel to avoid thermal stress; helium was used for the gas bearing, but only after passing it through a liquid nitrogen trap since otherwise the few parts per million of water vapor in cylinder helium were found to condense out in the bearing. Germanium lenses, cooled to liquid nitrogen temperature, were used to pass a collimated beam of light into the interferometer and to condense the exit beam onto the mercury-doped germanium detector.

Auguson and Young [40] have gone one stage further and have designed and constructed a liquid helium-cooled Michelson interferometer. The purpose of this instrument is to study extremely weak far-infrared emission from interstellar space, and ultimately extragalactic space. It is easy to realize that in order to detect such low intensity radiation, background emission from the instrument must be reduced to a minimum. At 100 μm, the radiant emittance from the interferometer at 77 K would be 6.4×10^{-7} watts/cm^2–cm^{-1} whereas at 5 K it would be 1.2×10^{-18} watts/cm^2–cm^{-1}, with an emissivity of one being assumed for each calculation. Cooling the interferometer with liquid nitrogen is insufficient for the detection of the very weak sources of interest, but below 5 K radiation from the spectrometer will not be restrictive on the measurements. However, cooling the interferometer places severe requirements on the interferometer alignment, the beamsplitter, and the drive mechanism.

Auguson and Young designed a four-sided support for the moving mirror that used strips and rollers to position the mirrors, which allowed parallel motion to be achieved within the specifications demanded by the highest frequency and desired resolution. It was later determined that "low micron size" tungsten diselenide powder makes a good cryogenic lubricant, and a linear bearing was constructed consisting of rows of ball bearings retained in a race surrounding a cylindrical shaft of hardened steel. After an extensive survey of possible beamsplitter materials, Mylar was found to be the optimum material for practical operation at low temperature.

The motor driving the moving mirror can be operated to accomplish two types of scan-modes: a repetitive rapid-scan mode through zero retardation for scanning an entire spectrum, or a scan-mode that avoids the zero retardation position so that the system can be used as a spectral line discriminator interferometer after the fashion of Foskett and Weinberg [41]. (These authors showed that a Michelson interferometer can be used as a detector of discrete line radiation while discriminating against continuum background radiation. Their method uses the fact that the Fourier transform of an emission line is a sinusoidal function which can be detected by means of a synchronous amplifier without using a chopper. If the interferometer mirror is displaced periodically, provided that the interferogram does not show great variations due to the background (i.e., provided that the mirror is well displaced from zero retardation) and provided that suitable filtering is carried out, this technique may be successfully used to discriminate weak or narrow band radiation from more intense background radiation.)

In view of the strong emission from the earth at low frequencies and the very strong absorbance of pure rotational lines of water vapor, this instrument has to be used above the earth's atmosphere, and has been designed to be flown on an Aerobee 150 rocket, since even at balloon altitudes, atmospheric radiation is many orders of magnitude greater than that from astronomical objects. This rocket only allows a total observation time of 5 min, but the sensitivity allowed by this liquid helium-cooled interferometer is sufficient to permit the desired measurements to be made.

REFERENCES

1. Beckman-RIIC Ltd., Eastfield Industrial Estate, Glenrothes, Fife, KY7 4NG, Scotland.
2. Sir Howard Grubb Parsons & Co. Ltd., Walkergate, Newcastle-upon-Tyne, NE6 2YB, England.
3. Societé Coderg, 15 Impasse Barbier, 92 Clichy, France.
4. Polytec Gmbh., 7501 Wettersbach-Karlsruhe, W. Germany.
5. L. Mertz, *Astron. J.*, **70**, 548 (1965).
6. Block Engineering, 19 Blackstone Street, Cambridge, Mass., 02139.
7. P. R. Griffiths, R. Curbelo, C. T. Foskett, and S. T. Dunn, *Anal. Inst. (Inst. Soc. Am.)* **8**, II–4 (1970).
8. Digilab Inc., 237 Putnam Avenue, Cambridge, Mass., 02139.
9. Idealab Inc., Union Street, Franklin, Mass., 02038.
10. EOCOM Corporation, 19722 Jamboree Road, Irvine, Calif., 92664.
11. Spectrotherm Corporation, 3040 Olcott Street, Santa Clara, Calif., 95051.
12. Willey Corporation, Box 670, Melbourne, Fla., 32901.
13. W. H. Steel, Aspen Int. Conf. on Fourier Spectrosc., 1970, G. A. Vanasse, A. T. Stair, and D. J. Baker, Eds., AFCRL–71–0019, p. 43.
14. J. Connes and P. Connes, *J. Opt. Soc. Am.*, **56**, 896 (1966).
15. M. Cuisinier and J. Pinard, *J. Phys.*, **28**, C2:97 (1967).
16. G. Guelachvili and J-P. Maillard, Aspen Int. Conf. on Fourier Spectrosc., 1970, G. A. Vanasse, A. T. Stair, and D. J. Baker Eds., AFCRL–71–0019, p. 151.

17. R. B. Sanderson and H. E. Scott, Aspen Int. Conf. on Fourier Spectrosc., 1970, G. A. Vanasse, A. T. Stair, and D. J. Baker, Eds., AFCRL–71–0019, p. 167.
18. R. A. Schindler, *Appl. Optics*, **9**, 301 (1970).
19. M. Born and E. Wolf, *Principles of Optics*, 2nd Edition, Pergamon Press, New York, 1964.
20. J. E. Chamberlain et al., *Infrared Phys.*, **6**, 195 (1966).
21. H. Sakai, Aspen Int. Conf. on Fourier Spectrosc., 1970, G. A. Vanasse, A. T. Stair, and D. J. Baker Eds., AFCRL–71–0019, p. 19.
22. R. J. Bell, *Introductory Fourier Transform Spectroscopy*, Academic Press, New York, 1972.
23. A. G. Tescher, Aspen Int. Conf. on Fourier Spectrosc., 1970, G. A. Vanasse, A. T. Stair, and D. J. Baker, Eds., AFCRL–71–0019, p. 225.
24. L. W. Thorpe, D. J. Neale, and G. C. Hayward, Aspen Int. Conf. on Fourier Spectrosc., 1970, G. A. Vanasse, A. T. Stair, and D. J. Baker, Eds., AFCRL–71–0019, p. 187.
25. P. Vogel and L. Genzel, *Infrared Phys.*, **4**, 257 (1964).
26. P. L. Richards, *J. Opt. Soc. Am.*, **54**, 1474 (1964).
27. J. D. Strong, *Concepts of Classical Optics*, Appendix F, coauthored by G. A. Vanasse, Freeman, San Francisco, 1958.
28. N. P. Hansen and J. Strong, Aspen Int. Conf. on Fourier Spectrosc., 1970, G. A. Vanasse, A. T. Stair, and D. J. Baker, Eds., AFCRL–71–0019, p. 215.
29. P. Bouchareine and P. Connes, *J. Phys. Radium*, **24**, 134 (1963).
30. A. Despain, F. Brown, A. Steed, and D. J. Baker, Aspen Int. Conf. on Fourier Spectrosc., 1970, G. A. Vanasse, A. T. Stair, and D. J. Baker, Eds., AFCRL–71–0019, p. 293.
31. M. Bottema and H. J. Bolle, Aspen Int. Conf. on Fourier Spectrosc., 1970, G. A. Vanasse, A. T. Stair, and D. J. Baker, Eds., AFCRL–71–0019, p. 210.
32. M. J. D. Low and F. K. Clancy, *Env. Sci. Technol.*, **1**, 73 (1967).
33. F. Leary, *Space/Aeronautics* (Sept. 1968).
34. A. T. Stair, Aspen Int. Conf. on Fourier Spectrosc., 1970, G. A. Vanasse, A. T. Stair, and D. J. Baker, Eds., AFCRL–71–0019, p. 127.
35. L. S. Block and A. S. Zachor, *Appl. Optics*, **3**, 209 (1964).
36. G. W. Ashley and A. G. Tescher, Aspen Int. Conf. on Fourier Spectrosc., 1970, G. A. Vanasse, A. T. Stair, and D. J. Baker, Eds., AFCRL–71–0019, p. 231.
37. R. A. Hanel et al., Aspen Int. Conf. on Fourier Spectrosc., 1970, G. A. Vanasse, A. T. Stair, and D. J. Baker, Eds. AFCRL–71–0019, p. 231.
38. J. Engel, G. Wijntjes, and A. Potter, Aspen Int. Conf. on Fourier Spectrosc., 1970, G. A. Vanasse, A. T. Stair, and D. J. Baker, Eds., AFCRL–71–0019, p. 289.
39. J. G. Moehlmann, J. T. Gleaves, J. W. Hudgens, and J. D. McDonald, *J. Chem. Phys.*, **60**, 4790 (1974).
40. G. C. Auguson and N. Young, Aspen Int. Conf. on Fourier Spectrosc., 1970, G. A. Vanasse, A. T. Stair, and D. J. Baker, Eds., AFCRL–71–0019, p. 281.
41. C. T. Foskett and J. M. Weinberg, *Appl. Optics*, **9**, 301 (1970).

5

AUXILIARY OPTICS FOR FTS

I. ABSORPTION AND REFLECTION SPECTROSCOPY

Infrared spectroscopy can conveniently be divided into two classifications: absorption and reflection spectroscopy and emission spectroscopy. In the former type of measurement, an infrared source emits continuous radiation over the bandwidth of interest, spectral information is generated (by a mono-chromator, an interferometer or any other device), and the radiation is sensed by a detector. The radiation is passed through (or reflected from) the sample of interest, which can be held anywhere between the source and the detector. The transmittance of the sample is calculated by taking the ratio of the energy transmitted through the sample at each frequency in the spectrum, $I_S(\bar{v})$, to the energy at the detector, $I_B(\bar{v})$, with either no sample or a reference cell in the beam.

$$T(v) = \frac{I_S(\bar{v})}{I_B(\bar{v})}. \tag{5.1}$$

The reflectance is equal to the ratio of the energy reflected off the sample to the energy reflected off a specular ($\sim 100\%$) reflector.

In a double-beam dispersive spectrometer, this ratio is usually determined sequentially for each frequency while the spectrum is being scanned, either using an optical null method or a digital ratiometer. Naturally, in Fourier transform spectroscopy, this type of ratio-recording cannot be performed since an interferogram contains information on all the spectral elements in the bandwidth of interest instead of just one. It is therefore the usual practice to measure all interferograms using single-beam techniques. The reference interferogram is measured first, the sample is then inserted in the beam, and the sample interferogram is measured. Each interferogram is then transformed separately, the ratio of the sample and reference spectra is calculated, and finally the spectrum is plotted out.

Most far-infrared interferometers operate in this single-beam mode and have only one position at which the sample can be held. As a result they are generally quite compact and easily able to be evacuated. The instruments manufactured by Grubb-Parsons [1] and Beckman-RIIC [2] are typical of this type of instrument. Radiation from a mercury lamp source is collimated, passed through the interferometer, focused onto the sample and refocused

onto the detector. The Grubb-Parsons Mark II interferometer uses polyethylene lenses for focusing which results in a fairly high energy loss at
frequencies above 200 cm^{-1}. The Beckman-RIIC FS–720 uses paraboloidal
mirrors to focus the beam into the sample and a polyethylene lens and
and light-pipe to refocus the radiation onto the detector. The optical layout
of this spectrometer is typical of several commercial and "home-made" far-
infrared spectrometers and is shown in Fig. 5.1. The sample compartment

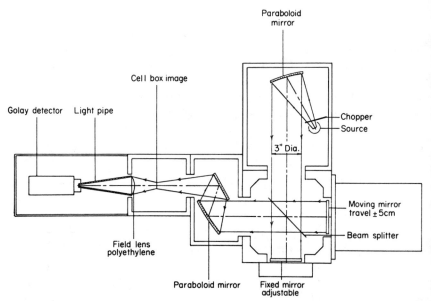

Fig. 5.1. Optical layout of the Beckman–RIIC FS–720 interferometer. (Courtesy of Beckman
Instruments Inc.)

of the instrument as shown here is designed for transmittance measurements
of large samples (discs, liquid cells, short path gas cells, etc.), but it can be
unbolted from the spectrometer and detector units and replaced by microsampling units, long path gas cells, cryostats or specular reflection attachments, with a provision for a polarizer where necessary. Between every
measurement the vacuum must be broken (at least in the cell compartment)
so that the sample can be changed.

The instruments made by Coderg [3] and Polytec [4] have more versatile
optical systems for absorption and reflectance spectroscopy in a vacuum. By
a suitable combination of mirrors, both of these spectrometers may be used
to measure the sample and reference interferograms without breaking the
vacuum in between the measurements. Figure 5.2 shows the optical layout

FIR 30 Optical Ray Diagram

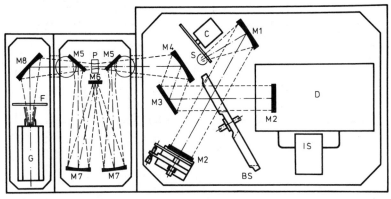

Fig. 5.2.   Optical ray diagram of the Polytec FIR–30 interferometer. Light from the source S is chopped at C and passed into the interferometer. The moving mirror, driven by the drive system D, is referenced by the Moiré fringe system IS. The beamsplitter is mounted on a wheel for rapid interchange of spectral regions. The radiation leaves the interferometer chamber and is passed into the sample chamber and then to the detector chamber where it is focused on the Golay detector D. (Courtesy of Polytec GmbH.)

of the Polytec FIR–30 while Figs. 5.3a and b show how the mirrors in the sample compartment can be positioned for the measurement of the transmittance and reflectance of solid samples and the transmittance of gaseous samples, respectively. For gases, the mirrors may be set so that by multiple reflection path-lengths of up to 5 m may be attained.

Thorpe et al. [5] have described a method of eliminating the need for measuring interferograms from the sample and reference beams sequentially. Their method allows the sample and background interferograms to be obtained simultaneously and under identical instrumental conditions (such as vacuum and source intensity). Their system involves a sum-and-difference technique whereby both beams are chopped at the same frequency so that energy is received from each beam alternately with complete obscuration between exposures (Fig. 5.4). The resultant detector output has one component at the chopping frequency proportional to the difference in amplitude between the two signals and another component at twice the chopping frequency proportional to the sum of the amplitudes of the two signals as shown in Fig. 5.5. The sample and background interferograms are separated from the two components numerically, transformed separately and ratioed. Spectra measured using this technique show considerably improved reproducibility over conventional sequential sampling methods because of the difficulty of maintaining of constant output from mercury lamp sources over long periods of time.

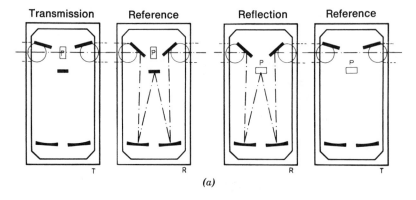

(a)

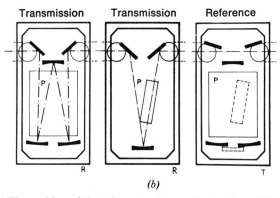

(b)

Fig. 5.3. (a) The position of the mirrors in the sample chamber of the FIR-30 for transmission and reflection studies of solids and liquids; (b) configuration of the mirrors in the sample chamber of the FIR-30 for transmission studies of gases. (Courtesy of Polytec GmbH.)

The only problem with this double-beam system is that is suffers from "cross-talk," that is, when one channel is closed and the other open, energy from the open channel is observed in the output signal representing the closed channel. The magnitude of the "cross-talk" is found to be dependent on the second harmonic component in the detector output, when the normal square-wave type of chopper is used. Hence the summing and differencing of the amplitude of the fundamental and second harmonic of the chopping frequency introduces false energy into both channels, with a resulting error in the measured spectrum. When a Golay detector is used for these measurements, the error due to cross-talk is small enough that the use of this technique still leads to noticeable improvements in the quality of the spectra.

This method of measuring sample and reference spectra during the same

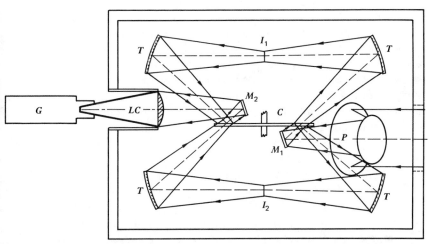

Fig. 5.4. Beam-switching unit designed to replace the beam-condensing optics of the Beckman–RIIC FS–720. The geometry of the chopper blade is such that the beam may be alternately transmitted and reflected to be re-imaged at positions I_1 and I_2. The beams follow paths of identical length to the Golay detector G. (Reproduced from [5] by permission of the author and Oriel Press; copyright ⓒ 1969.)

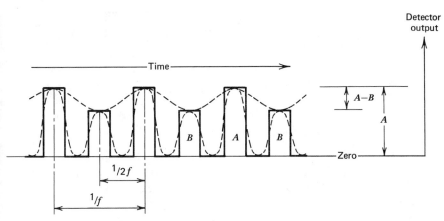

Fig. 5.5. Schematic representation of the signals in the "sum-and-difference" method of separating the signals from I_1 and I_2. (Reproduced from [5] by permission of the author and Oriel Press; copyright ⓒ 1969.)

scan was designed to minimize the effect of any drifting of the source energy. When a rapid-scanning interferometer is used for mid-infrared absorption spectroscopy, interferograms have to be measured sequentially for the sample and reference beams. The Digilab [6] FTS–14 spectrometer has a fairly simple means of achieving this end [7]. The optical layout of this instrument is shown in Fig. 5.6. An optional number of scans are signal-averaged in the reference beam after which the mirrors B and C are flipped so that the beam passes through the sample channel. The same number of scans are signal-averaged in this beam and the mirrors are switched again. The beam is switched between the sample and reference channels until the desired total number of scans are co-added. The number of individual scans co-added before the mirrors are flipped is determined by the rate of drifting of the instrumental conditions, in particular source intensity and dry air purge.

The optical system of this spectrometer is relatively simple, but the source unit is of particular interest. The size of the first focus, F_1, determines the solid angle of the beam passing through the interferometer, and whenever a resolution of 1.0 or 0.5 cm^{-1} is desired an aperture of 6 mm diameter is placed at this focus. Otherwise the aperture at this point (which is equal to the size of the focus F_2 in the sample compartment) has a diameter of 12 mm. This 12 mm aperture allows 2 cm^{-1} resolution to be achieved for all frequencies below 4000 cm^{-1}. The image size at the detector, F_3, is 2 mm, the same diameter as the detector itself, so that the system is "throughput matched" for mid-infrared spectroscopy at 2 cm^{-1} resolution, see Chapter 8, Section II.

An image size at F_2 greater than 12 mm in diameter is generally unrealistic

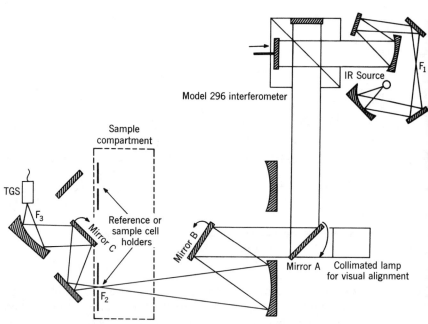

Fig. 5.6. Optical layout of the Digilab FTS–14 Fourier transform spectrometer. (Courtesy of Digilab, Inc.)

for mid-infrared absorption spectroscopy since most samples are as small or smaller than this. Thus the need to design a means of *increasing* the optical throughput of this interferometer for lower resolution work does not arise in practice.

The FTS–14 spectrometer has approximately the same beam dimensions in the sample compartment as most commercial grating spectrometers, and so can accept micro-sampling and reflectance accessories designed for conventional spectrometers with only a small reduction in efficiency. When such accessories are aligned on a conventional grating spectrometer, the visible light emitted from the infrared source is used to find the approximate position of all the mirrors in the accessory. This method cannot be used in a Fourier transform spectrometer with a germanium beamsplitter when the sample is held between the interferometer and the detector in view of the opacity of germanium to visible light. In spite of this disadvantage, holding the sample on the detector side of the interferometer rather than the source side is preferable for two reasons.

1. The sample is protected from much of the source emission and is therefore able to be held at a lower temperature.

2. Radiation emitted from hot samples is unmodulated and therefore undetected by the data system.

The optical system of the Digilab FTS–14 spectrometer has a source of collimated visible light which can be used in place of the infrared source by turning mirror A. After approximate alignment using this visible beam, the infrared beam is then switched back and the energy passing through the accessory is optimized by finding the maximum signal at zero retardation by repetitively scanning the interferometer over a short scan length. Thus unlike most of the far-infrared Fourier transform spectrometers, sample handling with this particular mid-infrared Fourier transform spectrometer is quite similar to that of conventional grating spectrometers.

A unique Fourier spectrometer which has been designed for the specific purpose of measuring diffuse and specular reflectance spectra is the Willey [8] Model 318 Total Reflectance Infrared Spectrophotometer. The optical layout of this system is shown in Fig. 5.7. The interferometer itself produces relatively low resolution spectra (>2 cm^{-1}), and its mirror position is continuously monitored by a helium-neon laser, as shown in the upper section of the diagram. The beam from the infrared source is modulated by the interferometer, and then passes to a rotating chopper which directs the beam to either of two paths. The chopper itself is divided into four equal sections. One quadrant is a specular mirror, and the opposite quadrant is removed so that the beam is transmitted. The other two quadrants are made of a

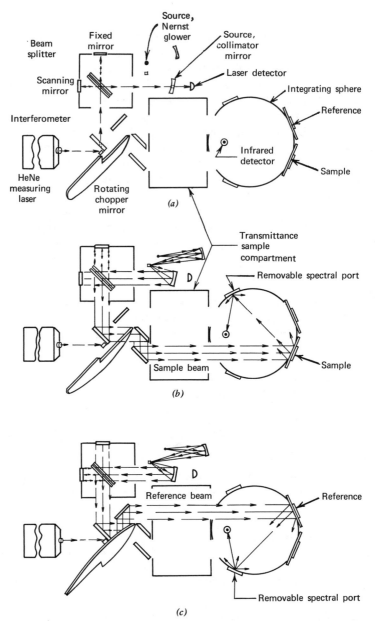

Fig. 5.7. Optical layout of the Willey Model 318 Total Reflectance Spectrometer. The retardation is monitored with a He-Ne laser in the fashion shown in (a). The chopper is divided into four quadrants, two of which are blackened and represent the zero energy reference. One quadrant is removed so that light falls onto the sample at the surface of an integrating sphere (b), while the other quadrant is a specular reflector which deflects the beam to the reference position (c). By removing the ports indicated on the walls of the integrating sphere, the specularly reflected components of the sample and reference beams will not reach the detector and only the diffusely reflected component is measured. (Courtesy of Willey Corporation.)

nonreflecting material which is known to closely approximate a black-body in the infrared. When the beam passes through the open portion of the chopper (Fig. 5.7b), it is then directed to the position where the sample under study is held. When the chopper rotates a full 180° from this position, the beam is reflected from the mirror quadrant of the chopper on to the sample held at the reference position (Fig. 5.7c). When the chopper rotates 90° from either the sample or reference position, the beam impinges on the black-body and is totally absorbed. The signal measured when the chopper is at this position gives the "zero" energy which is subtracted from the sample and reference signals so that real-time double-beam spectra may be obtained. The interferometer used for these measurements has to be of the slow-scanning type so that the sample, reference and zero energies are all measured at effectively the same retardation.

Both the sample and reference are held at the surface of an integrating sphere constructed with a diffusely reflecting gold surface. Using a flux-averaging device such as the integrating sphere, the collecting optics are equally efficient for measuring the reflection spectrum of a flat specular mirror or a highly scattering diffuse reflector. However, specularly reflected radiation from either the sample or the reference can be prevented from reaching the detector by removing ports on the wall where specularly reflected radiation from a flat sample would first hit the surface of the sphere. Thus both the total *and* the diffuse reflectance of any sample can be measured using this instrument. The capability of accurately measuring the total reflectance, even of rough samples, enables the optical properties of all types of materials to be readily calculated.

The integrating sphere used in this spectrometer has the advantage of collecting rays reflected at any angle, so that equal proportions of the rays reflected from the sample in all directions reach the detector. Unfortunately, the efficiency of the integrating sphere is, to a first approximation, proportional to the ratio of the area of the detector to the surface area of the sphere, so that for large integrating spheres with small detectors the device becomes rather inefficient. It was for this reason that, until the development of the interferometer, all measurements taken using an integrating sphere had to be of rather low resolution to allow a sufficient signal-to-noise ratio to be attained. The use of Fourier transform techniques represents a means of measuring a spectrum with a reasonable signal-to-noise ratio and resolution with good photometric accuracy.

Preliminary results using this instrument with a room temperature pyroelectric bolometer for measuring diffuse reflectance spectra are encouraging for low resolution work (20 cm^{-1}). However spectra measured at higher resolution show a rather high noise level. When a cooled mercury cadmium telluride photodetector or a mosaic pyroelectric detector is used instead of the single pyroelectric, spectra are measured with a sufficiently high signal-to-noise ratio to allow some very useful measurements to be made which have not previously been possible.

Another type of device used to measure specular and diffuse reflectance spectra is the ellipsoidal mirror reflectometer. The first such instrument where a Michelson interferometer was used to measure accurate reflectance spectra was described by Dunn [9]. Collimated light from a Michelson interferometer was passed through a port in an ellipsoidal mirror onto the sample which is held at one focus of the ellipsoid (Fig. 5.8). The light reflected from the sample was refocused at the infrared detector at the second focus

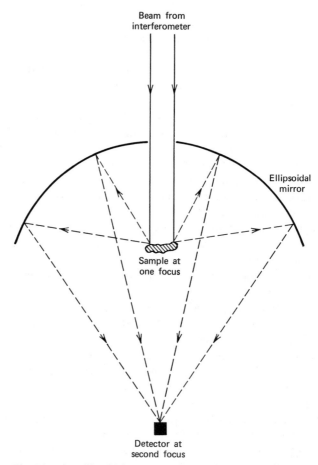

Fig. 5.8. An ellipsoidal reflectometer: radiation diffusely reflected from the sample is collected and focused onto the detector.

of the ellipsoid. Very accurate reflectivity values were able to be measured using this system.

II.   EMISSION SPECTROSCOPY

For practical purposes emission spectroscopy can be divided into two classifications: measurements where the source is in the laboratory close to the interferometer, and those where the source is remote and can be considered as being situated at infinity. In practice the size of remote sources can vary considerably: they can be extensive (as in the case of air-glow studies) or very small (for instance when the spectrum of a remote star is measured). An intermediate case is found when a fairly close (e.g., $\frac{1}{2}$ mile) source of one or two meters diameter is studied, such as smoke-stacks. Each type of measurement normally requires a telescope to collect the radiation; in the case of the extended sources the solid angle passing through the interferometer may be greater than the maximum allowed angle for the highest spectral frequency at the desired resolution. In this case an aperture stop must be placed at a focus somewhere between the telescope and the detector. For very small sources, the solid angle is generally well under the limiting value and no such aperture stop is needed. Laboratory experiments do not generally require a telescope, but rather require collecting optics of the type used to collimate source radiation in absorption spectrometers, and little need be said about this arrangement.

The most useful type of telescope used for remote studies is shown in Fig. 5.9. In this type of system, the half-angle of the beam passing through

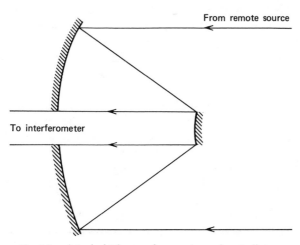

Fig. 5.9.   A typical telescope for remote sensing studies.

the interferometer is greater than the half-angle of the beam entering the telescope in the same ratio as the diameters of the primary and secondary mirrors of the telescope.

The method of ensuring that the detector is monitoring the source of interest varies depending on the type of measurement; for field measurements a small telescope can be mounted on top of the interferometer in such a way that its optical axis is parallel with that of the interferometer so that the system can be visually aligned. For moving sources, tracking techniques must be used. For example, in astronomy tracking usually requires a prior knowledge of the orbit of the body of interest. However, if a rapidly moving hot source, such as a rocket exhaust, is to be monitored, than a heat sensor can be used in the tracking system.

In emission spectroscopy of weak remote sources, radiation emitted from nearby sources (and especially the interferometer itself) can often be more intense than the radiation of interest. In the previous chapter two cryogenic interferometers were described where cooling had to be performed to reduce the level of the radiation emitted from the optics of the interferometer well below the level of that emitted from the source. Cooling the interferometer obviously presents severe problems and several other techniques for the compensation of background radiation have been described.

The simplest of these methods was developed for laboratory measurements of emissivity. Three spectra are needed to determine the emissivity of a sample accurately: the emitted energy from the sample, $I_S(\bar{\nu})$, the emitted energy from a reference black-body, $I_R(\bar{\nu})$, and the energy from a cooled black-body, $I_B(\bar{\nu})$. The spectrum $I_B(\bar{\nu})$ represents the background energy from the interferometer and must be subtracted from $I_S(\bar{\nu})$ and $I_R(\bar{\nu})$ to give the actual energy being emitted from the sample and reference. Thus the emissivity of the sample, $\varepsilon(\bar{\nu})$, can be calculated to a good approximation from the relationship:

$$\varepsilon(\bar{\nu}) = \frac{I_S(\bar{\nu}) - I_B(\bar{\nu})}{I_R(\bar{\nu}) - I_B(\bar{\nu})}. \tag{5.2}$$

A system has been described [10] where each spectrum can be measured sequentially, and this device is shown in Fig. 5.10. The top and the walls are cooled and coated with a material that is uniformly "black" in the infrared. The low temperature of the walls reduces their emission and the black surfaces minimize the amount of reflected light reaching the interferometer. Dry ice cooling is sufficient for measurements above 800 cm^{-1}, but if room temperature emissivity measurements are required below this frequency, cooling with liquid nitrogen becomes necessary. Any aperture around the sample must also be cooled or it must be a specular mirror in order that the only background viewed by the spectrometer is cold or black. A three-

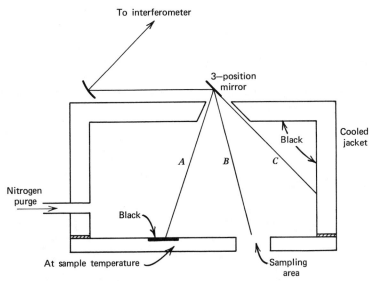

Fig. 5.10. The room-temperature emissivity attachment designed originally for measurements made with the Digilab FTS–14 spectrometer. Three signals are sampled in sequence: A, the signal from the reference blackbody at the same temperature as the sample, I_R; B, the sample signal, I_S; C, the signal from the cooled black walls, I_B.

position mirror at the exit aperture monitors the emitted radiation from the sample, reference and background, so that the emissivity can be calculated.

This device can be used when the sample is at or above room temperature, but for very weak sources even more care has to be taken, especially when the interferometer is in a mobile environment. The problem of measuring very weak emission spectra invariably arises for measurements taken in space. The IRIS spectrometer on the Nimbus III satellite [11] used a method rather similar to the one just described for room temperature emissivity measurements. A three-position mirror was oriented so that the interferometer monitored the Earth ($\equiv I_S$ in the previous example), an on-board reference black-body at about 280 K ($\equiv I_R$) or deep-space ($\equiv I_B$). In this way, a photometric accuracy of $\pm 1\%$ for terrestrial radiance was achieved.

Stair [12] has described a different method of distinguishing between radiation emitted from a remote weak source and radiation emitted from instrumental optics and windows. The source of interest in this study was an atmospheric emission spectrum measured from an aircraft at 12 km altitude. Calculations showed that this radiation would be less than radiation from the interferometer and the aircraft windows. A liquid nitrogen cooled chopper (situated outside the aircraft) was designed, and this is illustrated in Fig. 5.11. The chopper was painted black to obtain about 1% reflectivity. At 78 K, emission from the chopper becomes negligible compared to that from the source. However radiation reflected off the blade from the room temperature optics inside the aircraft was still larger than the terrestrial radiance. To reduce the intensity of the reflected radiation, a field-limiting cooled honeycomb is used to eliminate diffuse scatter from wide angles from the blackened surface of the chopper.

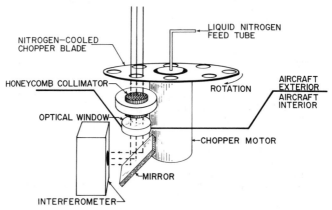

Fig. 5.11.   Schematic of the exterior liquid nitrogen-cooled chopper and field-limiting honeycomb of Stair. (Reproduced from [12] by permission of the author.)

When the chopper is at ambient temperature, the measured interferogram is typical of that from a black-body source. As the chopper blade is cooled, the magnitude of the interferogram at zero path difference decreases, indicating that most of the radiation being measured originates from the blade. A null point is reached when the atmospheric radiant energy is exactly equal to that from the chopper, since the two are exactly 180° out-of-phase. As the blade is cooled further, interferograms with much more structure indicative of the noncontinuum sources being studied are measured.

In view of the aircraft vibrations, Stair found it necessary to use a continuous scanning interferometer with a high scan speed, since it was found that this type of interferometer is the easiest to servo-control to offset the effect of vibrations. This arrangement cannot be defined as being a rapid-scanning interferometer under a strict definition since the modulation frequency of the radiation determined by the rotation frequency of the chopper must be higher than the product of the mirror velocity and the infrared frequency. With the eight-hole chopper shown in Fig. 5.11, the carrier frequency is about 1600 Hz, so that the signal frequency cannot exceed about 300 Hz. This restriction places an upper limit of about 0.1 mm/sec on the scan-speed of the interferometer.

REFERENCES

1. Sir Howard Grubb Parsons & Co. Ltd., Walkergate, Newcastle-upon-Tyne, NE6 2YB, England.
2. Beckman-RIIC Ltd., Eastfield Industrial Estate, Glenrothes, Fife, KY7 4NG, Scotland.
3. Societe Coderg, 15 Impasse Barbier, 92 Clichy, France.
4. Polytec Gmbh., 7501 Wettersbach-Karlsruhe, W. Germany.
5. L. W. Thorpe, R. C. Milward, G. C. Hayward, and J. D. Yewen, *Optical Instruments and Techniques, 1969*, Oriel Press, Newcastle-upon-Tyne, 1969.
6. Digilab Inc., 237 Putnam Ave., Cambridge, Mass., 02139
7. P. R. Griffiths, C. T. Foskett, and R. Curbelo, *Appl. Spectrosc. Revs.*, **6**, 31 (1972).
8. Willey Optical Corporation, Box 670, Melbourne, Fla., 32901.
9. S. T. Dunn, Ph.D. thesis, Oklahoma State University, 1965.
10. Digilab Inc. Room Temperature Emissivity Attachment for FTS–14 spectrometer.
11. R. A. Hanel et al., Aspen Int. Conf. on Fourier Spectrosc., 1970, G. A. Vanasse, A. T. Stair, and D. J. Baker, Eds. AFCRL–71–0019, p. 231.
12. A. T. Stair Jr., Aspen Int. Conf. on Fourier Spectrosc., 1970, G. A. Vanasse, A. T. Stair, and D. J. Baker, Eds., AFCRL–71–0019, p. 127.

DATA SYSTEMS

I. COMPUTATIONS ON A LARGE REMOTE COMPUTER

Until recently, the tremendous time saving gained through using an interferometer because of Fellgett's advantage had not been realized to its fullest extent because of the time delays involved in the computation of the spectrum from the interferogram. In the early days of interferometry, all interferograms had to be recorded, generally using a paper tape punch or magnetic tape recorder, and then carried to the computing center, read into the computer, and then transformed into spectra. This method not only had the disadvantage of requiring several visits to the computer center every week, but also frequently necessitated long waiting periods while other computer users had their jobs finished. Periods of as long as two or three weeks were known to this author before the computed data were returned to the spectroscopist. The increased speed of computers and the sophistication of modern time-sharing routines have meant that this type of time delay is now largely a thing of the past. However, it is still inconvenient and often costly to have to use a central computing facility for all spectral calculations.

Most of the early interferometers used by chemists were the slow-scanning type used for far-infrared spectroscopy. For these instruments, the data rate is sufficiently slow that a paper tape punch can be used to record the digitized interferogram. Indeed this is still the way in which data are recorded on most far-infrared Fourier transform spectrometers in current usage. Since the data rate is slow, time sharing on a large computer has also been successfully used with these interferometers, although it should be noted that this method does require the installation of a terminal close to the interferometer.

Before data processing can be initiated, a list of parameters must be read into the computer. These parameters include, for example, an identification number, the number of interferograms to be transformed, the sampling interval and resolution, the first and last frequency of the spectral plot, and information as whether the spectrum is to be used as the sample or reference in a ratio-recorded plot. The large computer has the advantage of being able to be easily programmed in a conversational language, such as FORTRAN, and programs for the FFT are available from several sources. The large computer can also be used to manipulate the data further, so that many operations can be carried out on the spectrum immediately after the FFT while the data are still in the computer memory.

Apart from time delays before the spectral data are returned, the use of a large remote computer has one other principal disadvantage. The computation step can frequently become rather expensive when time on a large machine has to be purchased. If many interferograms are being recorded each day, computing time for a year's operation can mount into a sizeable amount, and it is generally far preferable to use a minicomputer as a component of the spectrometer, so that the Fourier transform can be computed immediately after (or even during) the data collection step.

II.   SIGNAL-AVERAGERS FOR RAPID-SCANNING
INTERFEROMETERS

When rapid-scanning interferometers are used for Fourier transform spectroscopy, it is usually less economical to perform the Fourier transform on each interferogram individually than to signal-average interferograms and transform the result, since the only difference between successive interferograms should be noise. There is, of course, no difference between a spectrum produced from the transform of a signal-averaged interferogram and the resultant spectrum produced after individual interferograms are transformed and their spectra averaged, provided that nothing changes instrumentally between successive scans. The most important factor affecting the validity of signal-averaging is the necessity for coherent addition, that is, sampling each interferogram at precisely the same retardation values, and in particular the need to start data collection at the same point on the scan each time.

For interferometers with a white-light reference and a laser reference interferometer, coherent signal-averaging should take place automatically, but many early interferometers did not possess such fringe-referencing features. The effect of collecting the first data point at different retardations is illustrated diagrammatically in Fig. 6.1; two successive interferograms, A and B, are displaced by one sampling interval, so that the resulting interferogram, C, has a reduced intensity when compared with the resultant which would have been found if the error had not occurred (dotted line). The phase of the resultant is also intermediate between that of the interferograms A and B. Thus, not only does the signal-to-noise ratio of the averaged interferogram not increase as fast as if data collection had been initiated at the same retardation for each scan but also phase correction of the interferogram from a polychromatic source becomes quite difficult and poor intensity values may result in the spectrum.

If signal-averaging were carried out *correctly*, the signal-to-noise ratio of the signal-averaged interferogram should increase with the square root

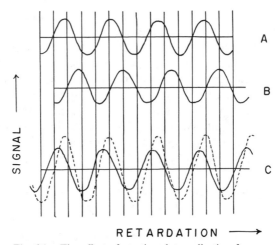

Fig. 6.1. The effect of starting data collection for successive sinusoidal interferograms, *A* and *B*, at different points. The two waves shown in *C* show the interferogram (solid line) which is the resultant of *A* and *B*, compared with the one (broken line) which would have been measured had both interferograms been initiated at the same point as *A*. Each vertical line represents one sampling position.

of the number of scans, $\sqrt{N}$, as discussed in Chapter 8, Section I. The earliest interferometers used a constant frequency clock as trigger for each data point, the clock being started at the same time as the drive voltage for the moving mirror of the interferometer. If the velocity of the mirror drive varies slightly in such a system, the interferogram is not sampled at precisely the same retardation values during each scan, and the signal-to-noise ratio shows a dependence of $(N)^{1/x}$, where x is greater than two.

Many of these early rapid-scanning interferometers used hard-wired signal-averagers for data collection. For example, all the early Block Engineering [1] interferometers were equipped with either a specially-designed Block "Co-Adder" or a Fabri-Tek Inc. [2] Model 1062 BE signal processor into which the signal was fed. The Fabri-Tek processor could average up to 20,000 interferogram points per second if necessary. It contained up to 4096 18-bit memory units and its 10-bit analog-to-digital converter was sufficient for most applications for which the interferometer was used at that time.

At the end of the signal-averaging process, the interferogram could be punched out on a paper tape for processing on a large computer. If spectral data were required immediately after data collection at somewhat reduced accuracy than if a digital Fourier transform were performed, the interfero-gram could also be analyzed with an audio-frequency wave-analyzer, some of which are described in the next section. Several low energy measurements were performed using these interferometers in a field environment, where the interferogram and clock signals were recorded on two-track magnetic tape. After the measurement was completed, the tape recordings were played back to the signal-averager in the laboratory, so the amount of equipment required for field measurements was quite small.

III. ANALOG WAVE-ANALYZERS

Audio-frequency wave-analyzers can be used to determine the amplitude of each frequency component in the interferogram. The analog interferogram is rapidly and repeatedly presented to the wave-analyzer at such a rate that the component frequencies in the interferogram fall in the audio-frequency range. The wave-analyzer is slowly tuned through the audio-frequency spec-trum, and a signal is produced which can conveniently be recorded on a X-Y plotter. A plot of intensity versus audio-frequency results and since there is a linear relationship between the infrared and audio-frequencies, the output of the wave-analyzer is a single-beam spectrum, see Fig. 6.2.

This technique represents the cheapest method of transforming the inter-ferogram to a spectrum, but at the cost of a certain degree of accuracy, sensitivity, and flexibility.

A commercially available unit for slow-scanning interferometers which uses an audio-frequency wave-analyzer is the FTC–100 Fourier transform computer manufactured by Beckman-RIIC [3, 4]. Analog information from the amplifier is digitized and stored serially in a ferrite core matrix memory as binary words of 12 bits. Two sections of 1024-word capacity are available to enable a ratio of two spectra to be obtained as the two interferograms are simultaneously analyzed. Computation consists of cycling the store, con-verting back to an analog waveform and passing this output waveform through the wave analyzer. The program provides for either section of the store to be transformed and displayed separately, and the ratio of the two to be displayed to eliminate background effects. The pen carriage is coupled directly to the timing system so that the frequency accuracy is independent of the operator. A diagram of the FTC–100 computer in combination with the FS–620 interferometer is shown in Fig. 6.3.

The same manufacturer has gone one stage further with their FTC–300 system. Here the interferogram is stored in a 20K-word memory, so that ratio-recorded spectra are able to be computed at higher resolution or over a greater frequency range than was possible using the FTC–100 computer.

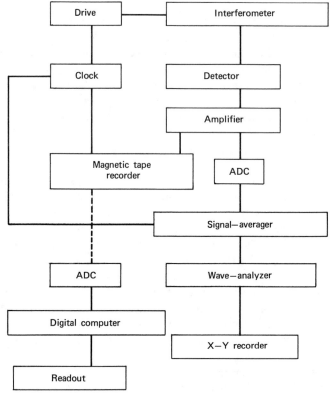

Fig. 6.2. Block diagram of an early type of Fourier transform spectrometer in which the interferogram and clock signals could could be recorded on two-track tape for later computation, or signal averaged and transformed directly using an audio-frequency wave-analyzer.

IV. USE OF MINICOMPUTERS

The use of minicomputers as components of a dedicated data system has revolutionized infrared Fourier transform spectroscopy more than any other single factor. Minicomputers give several important advantages over all the other systems described earlier in this section.

1. By attaching peripheral equipment such as disc files or digital magnetic tape units, the storage capacity can be increased essentially to any desired amount.

2. Recording data directly into the core memory of a minicomputer or onto discs or magnetic tapes is much more reliable than the use of a paper-

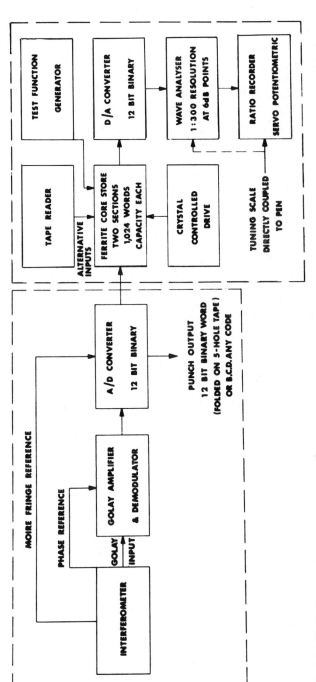

Fig. 6.3. Block diagram of the combination of the Beckman–RIIC FS–620 interferometer and the FTC–100 data system. (Courtesy of Beckman Instruments.)

tape punch. Paper-tape punches can easily get jammed and "miss" a bit with the result that a sine wave is generated across the complete spectrum (see Chapter 2, Section III).

3. Since the minicomputer is programmable it is very much more flexible than a hard-wired signal-averager. The principal advantage of this flexibility is found in the fact that a digital Fourier transform can be performed immediately after data collection has been completed or even in the "real-time" mode, while data are still being collected, so that an almost immediate presentation of the spectrum is available.

4. No information is lost from the interferogram during a digital transform so that the spectrum is more accurate than spectra derived using a wave-analyzer.

5. The computer may be used for controlling the spectrometer during data collection. For example, mirrors on the optical bench of the spectrometer can be positioned so that the beam passes through the desired path (e.g., sample or reference in a double-beam system). Sensing lines can be fed back to the computer to check on failure conditions (such as the reference laser not being sufficiently intense).

6. After the spectrum has been computed, the data may be readily processed further; several of these operations are described later in this section.

Many of the advantages of the dedicated data system will be illustrated by routines written for the disc data system used in Digilab's Fourier transform spectrometers [5], which is certainly the most versatile data system commercially available for infrared Fourier transform spectroscopy at this time. A block diagram of the Digilab FTS–14D spectrometer is shown in Fig. 6.4, in which the data system appears to be fairly simple. In practice, however, the spectrometer turns out to be rather complex electronically due to all the control functions that are carried out by the computer. For example, during data collection the computer regularly:

(a) checks the air pressure in the bearing of the interferometer drive mechanism;

(b) checks the intensity of the laser reference signal;

(c) checks the position of the mirror allowing light from the visual alignment source to pass through the sample compartment (see Chapter 5, Section I);

(d) checks that the beam is passing to the correct channel (front or back beam) to the detector;

(e) retraces the moving mirror of the interferometer after the required number of data points have been collected.

The failure conditions (a, b, and c) are monitored through the computer

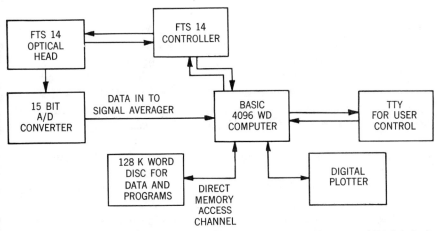

Fig. 6.4.   Block diagram of the Digilab FTS–14D spectrometer. (Courtesy of Digilab Inc.)

interrupt system. The spectrometer conditions (d and e) are controlled via programmable control and status registers. Data may be collected either through programmable registers or through the data channel facility. Data and programs are transferred between core memory and the auxiliary memory (disc) using both the computer interrupt system and the data channel, while data is acquired in real-time in a closed-loop program using the interferometer, disc, and core memory.

Thus it can be seen that the input/output (I/O) structure of the computer provides the key to data acquisition and spectrometer status and control. Most data systems have a teletypewriter, digital plotter, analog-to-digital converter, control and status registers and, often, a hardware multiply/divide unit on the I/O to speed the FFT. Many other types of peripheral equipment can also be interfaced to the minicomputer, such as magnetic tape units for additional storage of data and oscilloscopes for displaying spectra or interferograms before hard copy is made on the (slow) digital plotter.

On most Fourier transform spectrometers which have been designed for chemical spectroscopy, interferograms may be measured either single-beam or from two separate beams sequentially, in which case either the front or the back beam may be specified as the sample beam. Before data collection is initiated, the resolution and the number of scans in the sample and reference beams must be specified, and apodization and phase correction parameters are entered before the computation of the spectrum from the interferogram. In most Fourier transform spectrometers, all the parameters for data collection, computation, and plotting are entered before the measurement, and the complete operation of measuring the spectrum is initiated with one command.

The plotting routines of the Digilab data system illustrate how a dedicated computer can be used to effectively replace the knobs of a conventional analog spectrometer while, at the same time, giving the Fourier transform spectrometer increased capabilities over an analog instrument. Several of the plot parameters control routine functions such as the starting and final frequencies of the output, ordinate scale expansion and abscissa scale expansion. Others are used to give increased flexibility such as the capability of controlling the height of the ordinate scale or being able to expand the data in the frequency range under investigation automatically to fill this range if desired. For a ratio-recorded spectrum, the type of output (linear transmittance, linear absorbance or linear log absorbance) may be controlled, and the maximum and minimum values of the ordinate scale can be changed to be compatible with any of these plotting modes.

Whereas background effects in absorption and reflection spectroscopy are eliminating by ratioing, in emission spectroscopy they are eliminated by *subtracting* the background spectrum from the sample spectrum. The Digilab data system has a very versatile routine for emission spectroscopy whereby a background spectrum can be scaled and then subtracted from the sample spectrum. This routine has several useful applications in chemical spectroscopy. If the emission spectrum of a thin layer of a material on an emitting substrate is to be measured, the spectrum of a "clean" sample of the substrate material at the same temperature as the sample can be subtracted from the sample spectrum. If a cooled detector is being used, the background from the interferometer at or near room temperature may be eliminated in the same way. If emissivity measurements are to be made using the method described by Eq. 5.2 in the previous chapter, both subtracting and ratioing operations are necessary and can be carried out in any order using this data system.

Data systems which have a disc interfaced to the minicomputer usually use at least one of the tracks on the disc for the storage of programs. These program tracks are equipped with a "read-only" function so that no spectral data can be accidentally written over the programs. Because of the length of the complete programs required for this type of versatile data system, only small sections of the total program are able to be used in the core memory of the minicomputer at any one time. The remaining tracks on the disc are used for the storage of data either in the form of interferograms or spectra. In view of the nature of the Cooley-Tukey FFT algorithm, the size of the interferogram array should be 2^N, where N is an integer. Thus, interferograms of 1K, 2K, and 4K data points are usually collected and transformed on most data systems (including all-core systems), while longer transforms (8K, 16K, 32K, etc.) can be performed with disc data systems. The computational method used requires that a space equal to the size of the interferogram is available for the storage of the real (cosine transform) and imaginary (sine transform) portions of the spectrum after the FFT, each of which has the same length as the original interferogram. Therefore, for example, with a 128K-word disc on which some space is used for storing the programs, a 64K-word interferogram could be signal-averaged, but there would be insufficient room for the transform. On the other hand, there is sufficient space for two 32K interferograms to be stored and transformed sequentially provided that the space used for the imaginary part of the first spectrum can be used for the imaginary part of the second spectrum, that is, provided that only the real part of each spectrum is of interest.

Generally much smaller interferograms than 32K are collected in chemical spectroscopy. For example, a spectrum measured at 8 cm^{-1} resolution gives sufficient information to solve many problems by mid-infrared spectroscopy, and a 1 K word interferogram transforms into a spectrum of this resolution with a folding frequency close to 4000 cm^{-1} (see Chapter 2, Section I). Thus even for double-beam spectra at this resolution, only 3K words of storage are needed (1K each for the real sample and reference spectra and 1K for the imaginary part of the final spectrum to be computed). The remaining memory locations in the disc can be used to store other previously measured spectra. These spectra can then be recalled at any time for further processing or replotting.

Special purpose programs have been written on FTS data systems whereby stored spectral data can be processed to give additional information. For example, transmittance spectra are always calculated in FTS by taking the ratio of sample and reference energy at all frequencies in the spectrum, unlike double-beam grating spectrometers which generally use an optical-null technique. As the energy in the reference spectrum decreases, the "dead-pen" problem found with optical null spectrometers does not occur; instead, the noise level in the ratio-recorded transmittance spectrum increases, and if bands are seen in the reference spectrum which absorb more than, say, 99% of the energy over a certain frequency range, the transmittance spectrum over this region may become very noisy and little useful information will be able to be gathered from the spectrum over this range. Additionally, as the noise in a spectrum increases, plotting time on a digital recorder increases and the spectrum becomes less aesthetically pleasing. To get around this problem, a program has been written for at least one commercial Fourier transform spectrometer [5] whereby whenever the energy in the reference spectrum falls below a certain specified percentage of the maximum energy in the complete (single-beam) spectrum, the chart is automatically advanced to the next frequency at which the reference spectrum shows energy greater than this threshold value. This program has been found to be very useful when solution spectra are measured over frequency regions in which the solvent has several strong absorption bands and also in the study of atmospheric spectra over very long path-lengths, where lines due to water vapor can show very high absorbances and can mask much of the spectrum between 1200 and 2000 cm^{-1}. By performing the measurements at high resolution to sharpen up these water lines, and then using this program so that only the window regions between the sharp lines are plotted, good data on weak bands occurring in this region of the spectrum can be obtained.

Another program that has been written for an FTS data system is also of importance for chemists measuring the spectra of solutions. This program involves the elimination of the effect of strong absorption bands of solvents on the transmittance spectrum of a solution when the solvent reference is present at a different thickness in the sample and reference cells. Even when identical cells are used to hold the sample and reference liquids, displacement of the solvent by the solute will cause "negative-going" bands due to the solvent to be seen in the absorbance spectrum in regions where the solute has no absorption bands. The method by which such features can be compensated involves storing the absorbance $(-\log_{10} T)$ spectrum of the solution, $A_{(S+R)}$, and that of the solvent reference, A_R, both spectra being measured with reference to an empty reference beam. If the solvent has precisely the same optical thickness in both cells, then subtraction of A_R from $A_{(S+R)}$ would yield the absorbance spectrum of the solute A_S. However if the solvent bands are not completely compensated in this way, A_R may be multiplied by a factor x (where x is chosen using a process of trial-and-error by the operator) [5] so that when xA_R is subtracted from $A_{(S+R)}$, a spectrum with no trace of the solvent bands results. Another method which has been written to achieve the same goal [6] involves selecting a frequency at which the solvent has an absorption band but at which the solute is known not to absorb. The factor x is then computed automatically so that the resulting absorbance is set to zero after the subtraction.

A further example of the type of secondary data processing that can be performed on these data systems was encountered during a project designed to measure the optical characteristics of some narrow band-pass filters for for which the centroid of area between certain defined frequencies was of importance. Here it was a fairly simple matter for a program to be written to compute the centroid from the spectral data stored in the memory of the data system and to print it out after plotting the spectrum. Similar programs may be written to determine the integrated absorbance of spectral bands measured in quantitative analysis.

Another important spectroscopic application for analytical chemists which has been made possible by the presence of a flexible data system concerns the interface of a rapid-scanning interferometer with a gas chromatograph

(GC–IR). This combination has for several years been suggested as one of the principal applications for FTS in analytical chemistry, since it makes it possible to measure the infrared spectra of GC peaks as they elute from the chromatograph without trapping the sample. Although the feasibility of these measurements was shown several times using a signal-averager/wave-analyzer combination, the technique only became routine with the development of the software for collecting data during the time that the sample represented by each GC peak is at its optimum concentration in the infrared beam, and for storing these data immediately after the measurement for subsequent plotting and further detailed analysis.

For a typical GC–IR system, the effluent stream from a GC is passed into a heated light-pipe gas-cell via a heated transfer line. The signal at the GC detector is monitored, and when it reaches a certain threshold voltage, the status (off/on) line to the computer changes level. After a preset delay of t sec, to allow the sample to pass from the GC detector to the gas-cell, data collection is initiated and continued until t sec after the status line has returned to its original level.

Two methods of data collection and processing are currently in use for GC–IR. The most common technique [5, 7] involves signal-averaging inter-ferograms while the concentration of each GC peak is above the threshold level in the gas-cell. When the concentration falls below this level, the averaged interferogram is transferred to a given storage area on the disc. The same procedure is followed for subsequent GC peaks, with each inter-ferogram being stored in sequential filing areas on the disc. At the end of the chromatogram, each array is transformed and the spectra are ratioed against a previously measured reference spectrum taken through the empty gas-cell.

In the second method [8], the interferogram from the current scan is collected *at the same time* as the interferogram from the previous scan is being transformed. Thus in this case, *spectra* rather than interferograms are averaged and stored. With such a system, spectra can be viewed on oscilloscope display a very short time (~ 2 sec) after a given interferogram has been measured, thereby allowing the chemist to interact with the data system in approximately real time. The operator has the chance, therefore, to recognize peaks which have not been resolved by the chromatograph, and to collect and store the data accordingly. A further advantage of this computational technique is that since spectra, rather than interferograms, are stored, more information can be stored on a given disc, since it takes less room to store a spectrum than the corresponding interferogram. A disadvantage of the technique is that it necessitates more *core* in the mini-computer than the method in which interferograms are collected and stored.

The delay time, t, may be measured before the start of the chromatogram and remains constant for a given carrier gas flow rate. To measure t, a small sample of a strong infrared absorber is injected into the chromatograph under such conditions that a single *sharp* peak is measured on the GC recorder. The time at which the GC peak reaches a maximum value is monitored electronically, and starting from this moment the interferometer is set repeatedly scanning (at approximately one second intervals) during which time the amplitude of the interferogram at zero retardation is monitored. When the sample is present at its highest concentration in the gas-cell, the absorption of the IR beam is a maximum; thus the amplitude of the interferogram at zero retardation is a minimum.* The time between the maximum signal at the GC detector and the minimum amplitude of the interferogram at zero retardation gives the delay time, t.

For the conventional method in which GC–IR measurements are made, only one beam path is necessary.† However in view of the weakness of most of the absorption bands being measured, it is necessary to have a very flat baseline in the transmittance spectrum so that large ordinate expansions can be performed. To achieve this objective a reference spectrum of the empty gas-cell is measured before the sample is injected into the chromatograph. The reference interferogram is generally measured using at least four times the number of scans expected for the widest GC peak in the run; in this way the noise contribution is kept as low as possible and most of the noise in the transmittance spectrum is contributed by the (time-limited) sample measurement.

The technique of measuring the sample and reference spectra for a ratio-recorded plot with different numbers of scans is also commonly used when the same background spectrum is used as reference in the calculation of the transmittance spectra of several samples. For example, in measuring the mid-infrared spectra of microsamples [9], a low noise background spectrum should be measured through a reference cell in a beam-condenser and used for all sample spectra measured subsequently during the same day. If this reference spectrum is measured using, say, 1000 interferogram scans, and if each sample interferogram is then measured using only 250 scans (which usually takes about five minutes at low resolution) the measurement time is kept to a minimum. If the same spectrum were measured using a double-beam technique, it would have taken approximately seven minutes to run *each* of the sample and reference interferograms to produce a transmittance spectrum with the same noise level. Any time that more than two spectra

---

* For GC–IR systems which use the dual-beam technique described in Chapter 7, Section I, the interferogram signal at zero retardation would be a maximum when the sample is at its maximum concentration.

† When the dual-beam technique is used, two identical gas-cells are needed for the effluent from the sample and reference columns of the chromatograph.

are to be run with the same reference, this technique will result in considerable time savings.

On the other hand, if the sample transmits only a very small percentage of the incident energy while the background for the transmittance spectrum is the unattenuated instrumental background, it is profitable to use far fewer scans for the reference than the sample interferogram. For example, if the average background transmittance of a sample is 1%, it requires 10^4 scans to achieve the same signal-to-noise ratio in a single-beam spectrum of the sample as it would take using one scan through the reference beam. Thus it is obviously unnecessary to signal-average more than a very few scans of the reference interferogram in order that effectively all the noise in the transmittance spectrum is contributed by the single-beam spectrum of the sample.

Although it is certainly true that some of the features discussed above are equally applicable to data processing using a remote computer rather than an on-line minicomputer, the ease of operation and the immediate nature of the result make the dedicated computer data system a highly desirable component of a modern Fourier transform spectrometer. The time savings in data processing and the capability of the operator to interact with the data system certainly account in no small measure for the rapidly increasing popularity of infrared Fourier transform spectroscopy.

V. "REAL-TIME" DATA SYSTEMS FOR FTS

Data systems of the type described above have been designed to collect and store the digitized interferogram first, and to compute the Fourier transform using the FFT algorithm at the end of the data collection step. It was noted in Chapter 3, Section IV that the technique of "real-time" computation could be used to give a readout of the spectrum after each interferogram point has been collected. This technique has now been used with several slow-scanning interferometers, and recent advances which will be described at the end of this section suggest that real-time computational techniques might be used progressively more in FTS.

The first real-time data systems to be described in the literature did not use general-purpose computers. The first such system was described by Yoshinaga et al. [10] in 1966 and used a special purpose digital computer. In 1969, Hoffman [11] described an analog data system for this purpose. However, the real-time computing technique did not appear to be of great importance until quite recently when the first real-time data system using a general-purpose minicomputer became commercially available.

The system was described by Levy et al. [12] and forms the basis of the Coderg [13] Fourierspec 2000 Fourier transform spectrometer. A block

DIGITAL COMPUTER   INTERFACE   PERIPHERY

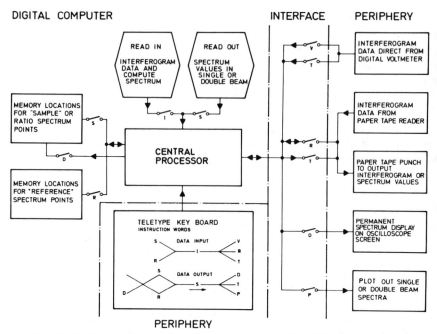

PERIPHERY

Fig. 6.5. Block diagram of the data system of the Polytec FIR–30 Fourier transform spectrometer on which real-time transforms are performed. (Courtesy of Polytec GmbH.)

diagram of the computing system of a similar spectrometer is shown in Fig. 6.5. The output of the detector, modulated by the chopper, is amplified, synchronously rectified and smoothed, and fed to a digital voltmeter. In order to remove the d.c. component of the interferogram and to increase the dynamic range of the system by one bit, the electrical interferogram is backed-off by a constant voltage so that the average value of the interferogram at large retardation is set as precisely as possible at zero volts. On receipt of a Moiré command pulse, the signal is digitized and fed into the computer. The interferogram values can be simultaneously punched out on paper tape in case computation of the full spectrum is desired on a larger computer at the end of the measurement. This procedure is necessary when high resolution data are to be examined over a wide frequency range.

The computational method described in Chapter 3, Section IV is summarized in Fig. 6.6. The computer used by Levy et al. is a Varian 620i minicomputer with 4K of core. With this memory size, up to 1500 frequency points are able to be computed for a single-beam spectrum, or 750 frequency points if the ratio between two spectra is being found. The remaining locations are used for programs and instructions (1572 words) and a table of cosine values (1024 words).

The time of computation of this system, T, is found to be almost directly

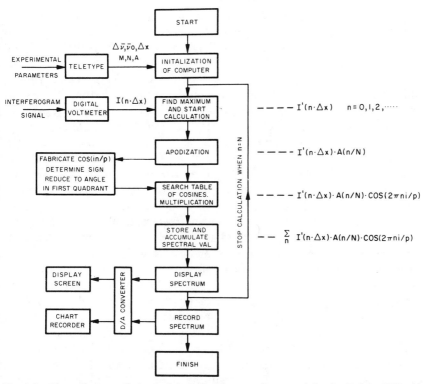

Fig. 6.6. Flow diagram of the real-time Fourier transform used in the Polytec FIR–30 spectrometer. Δx is the sampling interval, $n\Delta x$ is the retardation, N is the total number of points, $\Delta \bar{\nu}$ is the frequency interval of the output, $p = (\Delta \bar{\nu} \cdot \Delta x)^{-1}$, and i takes a range of integral values of $\bar{\nu}_0/\Delta \bar{\nu}$, $\bar{\nu}_0/\Delta \bar{\nu} + 1, \ldots$ and so on. (Courtesy of Polytec GmbH.)

proportional to the number of frequency points, M, calculated. T is given to a very good approximation by the formula:

$$T = 0.05 + 1.2 \times 10^{-3} \, M \text{ sec}$$

so that the time required to calculate 1500 frequency points is a little under two seconds. Thus it is apparent that the scan-speed of the interferometer must be quite slow if the full 1500 frequency points are to be calculated. However, if only a limited frequency range is of interest the speed of this data system places few restrictions on the operating speed of a typical far-infrared interferometer.

Using the Coderg data system, no phase correction can be carried out; as a result, the first sampling point must be set accurately at zero retardation. The photometric accuracy of spectra measured using the interferometer/real-time computer system described above is determined principally by the precision with which the interferometer can be set at zero retardation. To obtain less than 1% distortion, it is necessary for the sampling error, ε, to be less than 1% of the sampling interval. Levy et al. consider this to be the principal drawback of the method. They consider it unlikely that phase correction procedures (which must necessarily take place after several points after the zero retardation point) could be carried be carried out without endangering the speed of the real-time method of computation or substantially reducing the number of frequency points that can be calculated.

However, the designers of a second commercially available far-infrared Fourier transform spectrometer which also uses a real-time computing technique, the Polytec FIR–30 [7], claim to have overcome this problem. In this system, the position of the fixed mirror can be changed to coincide with the zero retardation point by means of a very fine translational adjustment. Any residual phase errors are then computed by fitting a parabola to the largest three sampled points around zero retardation. The interferogram retardation values are then adjusted accordingly. Descriptions of this spectrometer do not mention whether the phase-correction procedure is carried out during the measurement or before it. If it is performed during the measurement, the arguments of Levy et al. [12] still apply; if carried out before the measurement, then the interferometer has to be rather stable over a short period of time so that the moving mirror can be retraced over zero retardation after the short phase measurement, and the full measurement scan started.

The data system of the Polytec FIR–30 shows several advances over the earlier Coderg system. A 4K-word Data General Nova 1200 computer is used which is considerably faster than the Varian 620i. This computer has enabled the time for computing 1500 output frequency points to be reduced to 0.4 sec, which is within the typical data rate of a slow-scanning far-infrared interferometer. The amount of storage space for output points has also been increased in this data system so that 3000 points of a single-beam spectrum may be computed using the 4K computer or a 1500 point ratio-recorded spectrum may be computed. All calculations can be carried out in double-precision (32 bits) so that the residual errors due to round-off in the computation are negligible.

The two real-time computing systems just described were designed around slow-scanning, far-infrared interferometers with a slow data rate and relatively short frequency range. It is obvious that the real time computing method will show less compatibility with rapid-scanning interferometers where signal-averaging of interferograms is used to increase the final signal-to-noise ratio. However, several very high resolution interferometers have been described for mid-infrared spectroscopy where signal-averaging techniques are not applied. For these interferometers up to 10^6 data points may be collected at a relatively high data rate (> 10 points/sec). In view of their wide frequency range, high resolution and high data rate, it would seem on first inspection that the real-time computing technique would have very little application for this type of measurement. However, it is just in this field that some of the most significant advances in real-time computing techniques are being made. For high resolution measurements where one scan can take several hours to complete, it is particularly important to check whether the spectral bands of interest to the operator are in fact being measured during that measurement. If they are not, the run can be aborted before several further hours of measurement are wasted. By only looking at a very small fraction of the total frequency range, *but in a region in which important spectral bands are known to absorb*, Connes and Michel [14] have developed a method of computing 1000 output points at the rate of 50 input points/sec. This represents a gain in speed of almost a factor of 20 over the Polytec data system described previously.

The data system that was used to achieve this increase involves no radically new principle. It is simply a digital computer with a hard-wired classical Fourier transform algorithm. Analog devices are used only for apodization and at the output for interpolation between spectral points to present a continuous curve on the oscilloscope. The two features of this data system which differ from general purpose computers are the elimination of all programming by the use of hard-wiring and storing the spectrum in a sequential-access circulating memory (which also happens to be cheaper than a random access memory). It is envisioned [14] that this system will ultimately be able to output 20,000 frequency points at the rate of 50 input points/sec. According to Connes and Michel, even this performance could be improved upon. Using only commercially available integrated circuits and memories, a factor of at least three, and possibly ten, improvement in speed could be realized. The limit of 20,000 words output corresponds to the economic limit of the magnetostrictive delay line storage used in this system. Above this limit, drum or disk storage could be used.

Connes and Michel stress the general application of this type of data system to all types of interferometers except those designed for the spectroscopy of high intensity, short lifetime sources. However, the use of real-time techniques with rapid-scanning signal-averaging interferometers still does not appear to be beneficial at this time. Although it is true that this type of computer should be cheaper to build than the equivalent FFT computer because of both the simplicity of the addressing scheme and the elimination of large random access memories, two other features have to be taken into account. Since these are hard-wired computers, many of the attractive features of general-purpose minicomputers described in the previous section cannot be utilized. Also, at the present time, general-purpose minicomputers are in large-scale production and their price is actually being reduced annually, the special-purpose computers of the type envisioned by Connes would only be produced in limited quantity, so that their price would probably be higher in spite of their simplicity.

Therefore, in summary, it is unlikely that real-time Fourier transform spectroscopy will be used to any great extent for mid-infrared *chemical* spectroscopy where rapid-scanning interferometers show the greatest potential; however, for slow-scanning interferometers such as those used for far-infrared spectroscopy and ultrahigh resolution measurements in the mid-and near-infrared, real-time Fourier transforms may be applied to an increasing extent.

REFERENCES

1. Block Engineering Inc., 19 Blackstone Street, Cambridge, Mass., 02139.
2. *Now* Nicolet Instrument Corporation, 5225 Verona Road, Madison, Wisc., 53711.
3. Beckman-RIIC Ltd., Eastfield Industrial Estate, Glenrothes, Fife, KY7 4NG, Scotland.
4. J. N. A. Ridyard, *J. Phys.*, **28**, C2:62 (1967).
5. Digilab Inc., 237 Putnam Avenue, Cambridge, Mass., 02139.
6. Willey Corporation, Box 670, Melbourne, Fla., 32901.
7. Polytec Gmbh., 7501 Wettersbach-Karlsruhe, W. Germany.
8. Spectrotherm Corporation, 3040 Olcott Street, Santa Clara, Calif., 95051.
9. P. R. Griffiths and F. Block, *Appl. Spectrosc.*, **27**, 431 (1973).
10. H. Yoshinaga et al., *Appl. Optics*, **5**, 1159 (1966).
11. J. E. Hoffman, *Appl. Optics*, **8**, 323 (1969).
12. F. Levy, R. C. Milward, S. Bras and R. Letoullec, Aspen Int. Conf. on Fourier Spectrosc., 1970, G. A. Vanasse, A. T. Stair, and D. J. Baker, Eds., AFCRL–71–0019, p. 331.
13. Societé Coderg, 15 Impasse Barbier, 92 Clichy, France.
14. P. Connes and G. Michel, Aspen Int. Conf. on Fourier Spectrosc., 1970, G. A. Vanasse, A. T. Stair, and D. J. Baker, Eds., AFCRL–71–0019, p. 313.

DUAL-BEAM FOURIER TRANSFORM SPECTROSCOPY

I. REDUCTION OF DYNAMIC RANGE

One of the more difficult measurements to make using interferometry is the detection of very weak absorption bands (less than 0.1%). There are several reasons accounting for this difficulty, of which the most important involves the dynamic range of the signal. In Chapter 2, Section II, it was shown that if an absorption band of half-width σ_s is to be distinguished from the background of half-width σ_b, the dynamic range, of the digitization system must be *at least* $\sigma_b/\alpha\sigma_s$, where α is the fraction of the radiation absorbed at the peak [1]. This is true if the band absorbs at the peak of the emission background, if not, then the dynamic range must be even higher.

Consider the following examples to determine the dynamic range required to detect a weak absorption band in the mid-infrared. For a band absorbing 0.1% having a half-width of 6 cm^{-1}, if $\sigma_b = 3000$ cm^{-1} the factor $\sigma_b/\alpha\sigma_s$ is equal to 5×10^5. Thus a dynamic range of 2^{19} is needed if this band is to be detected. This is greater than the range of linearity for most detectors and therefore either σ_b has to be reduced using an optical filter (thereby simultaneously reducing Fellgett's advantage) or signal-averaging techniques must be used. Even using a signal-averaging system where the signal-to-noise ratio per scan is 2^{14}, it would still take at least 1000 scans, $\{(2^{19}/2^{14})^2\}$, to attain this signal-to-noise ratio. Thus a rapid-scanning interferometer can be used with an incandescent source and bolometer detector to detect a band of this strength, provided that the signal-averaging and FFT are carried out in double-precision in a normal data system.

The signal-to-noise ratio in the interferogram could be raised by between one and two orders of magnitude by substituting a sensitive photodetector for the bolometer, but under these conditions both the linear range of the detector and the dynamic range of the ADC would usually be exceeded. Under these circumstances it is far preferable to attempt to reconfigure the experimental arrangement to yield stronger absorption bands at the cost of the total energy flux at the detector.

An example of this is found in infrared microsampling [2] using a cooled mercury-cadmium telluride detector. If the spectrum of a 1.5 mm diameter

171

KBr microdisc of a certain sample is to be measured under these circumstances, a signal-to-noise ratio is measured which exceeds the dynamic range of a 15-bit ADC when the normal scan rate encountered with rapid-scanning interferometers is used. In this case, the low noise level in the interferogram severely reduces the efficiency of the signal-averaging process. However, if the same quantity of material were used to prepare a 0.5 mm diameter microdisc, the area of the sample and hence the energy flux at the detector would be reduced by about a factor of 10. In this case it becomes possible to use a sensitive detector and still employ signal-averaging techniques. In addition, the absorbance of the bands becomes about 10 times greater which means that much less ordinate scale expansion has to be applied in order to observe the bands clearly.

Occasionally, however, it is not practicable to reconfigure the experiment, and under these conditions the technique known as dual-beam Fourier transform spectroscopy can be used to reduce the dynamic range of the interferogram measured with an intense source and a sensitive detector.

In its normal mode of operation the Michelson interferometer has one input beam of radiation and two output beams, one passing along an axis perpendicular to the input beam and the other passing back along the same path as the input beam. Due to the experimental difficulties involved in measuring the latter beam, only the beam passing in the direction perpendicular to the input beam is usually measured. In dual-beam interferometry, either two beams from identical sources are passed into each arm of the interferometer and one or both output beams are measured or both of the output beams from a single input beam are measured.

Let us consider the latter case, and denote the interferogram which does *not* return to the source as the *transmitted* interferogram, $I_t'(\delta)$, and the one which does return along the path to the source as the *reflected* interferogram, $I_r'(\delta)$. For an ideal beamsplitter, the two interference functions are given by the expressions.

$$I_t'(\delta) = \tfrac{1}{2} \int_0^\infty I(\bar{v}) \cdot d\bar{v} + \tfrac{1}{2} \int_0^\infty I(\bar{v}) \cos 2\pi\bar{v}\delta \cdot d\bar{v} \qquad (7.1)$$

and

$$I_r'(\delta) = \tfrac{1}{2} \int_0^\infty I(\bar{v}) \cdot d\bar{v} - \tfrac{1}{2} \int_0^\infty I(\bar{v}) \cos 2\pi\bar{v}\delta \cdot d\bar{v} \qquad (7.2)$$

Thus on addition:

$$I_t'(\delta) + I_r'(\delta) = \int_0^\infty I(\bar{v}) \cdot d\bar{v}. \qquad (7.3)$$

Stated verbally, Eq. 7.3 shows that the total energy output from both arms of the interferometer is equal to the energy input; that is, the Law of Conservation of Energy is preserved. On *subtraction* of 7.2 from 7.1 we obtain:

$$I_t'(\delta) - I_r'(\delta) = \int_0^\infty I(\bar{v}) \cos 2\pi\bar{v}\delta \cdot d\bar{v}. \tag{7.4}$$

This expression shows that amplitude of the modulated portion of the interferogram is twice as great when the "reflected" interferogram is subtracted from the "transmitted' interferogram than the amplitude of either interferogram alone. This principle has been used to increase the signal-to-noise ratio of spectra of weak sources, and will be discussed in Section II.

If a beam of radiation of intensity $I_1(\bar{v})$ is passed into one arm of a Michelson interferometer and a second beam, of intensity $I_2(\bar{v})$, is passed into the other arm, the interferogram of the output beam from the first arm is equal to $I_1'(\delta)$ where:

$$\begin{aligned}
I_1'(\delta) &= \tfrac{1}{2} \int_0^\infty I_2(\bar{v}) \cdot d\bar{v} + \tfrac{1}{2} \int_0^\infty I_2(\bar{v}) \cos 2\pi\bar{v}\delta \cdot d\bar{v} \\
&\quad + \tfrac{1}{2} \int_0^\infty I_1(\bar{v}) \cdot d\bar{v} - \tfrac{1}{2} \int_0^\infty I_1(\bar{v}) \cos 2\pi\bar{v}\delta \cdot d\bar{v} \\
&= \tfrac{1}{2} \int_0^\infty \left[I_2(\bar{v}) + I_1(\bar{v}) \right] \cdot d\bar{v} + \tfrac{1}{2} \int_0^\infty \left[I_2(\bar{v}) - I_1(\bar{v}) \right] \cos 2\pi\bar{v}\delta \cdot d\bar{v}. \tag{7.5}
\end{aligned}$$

If a rapid-scanning interferometer is used for this experiment, only the a.c. component in Eq. 7.5 would be measured. The transform of this component gives the *difference* between the intensities of the two beams at each frequency; consequently the technique is sometimes referred to as *optical subtraction*. When radiation from two identical sources is passed into each arm of the interferometer, no a.c. signal can be detected and this situation is known as an *optical null*. If there is only a very slight difference between the intensities of the two beams $I_1(\bar{v})$ and $I_2(\bar{v})$, the amplitude of the interferogram can be very small even though the amplitude of the interferogram from either source input individually may be very large.

If a sample having a transmittance $T(\bar{v})$ is held in one input beam to the interferometer and no sample is held in the other beam, the a.c. interferogram measured if radiation from identical sources is passed into each arm of the interferometer is given by

$$I(\delta) = \tfrac{1}{2} \int_0^\infty I(\bar{v})[1 - T(\bar{v})] \cos 2\pi\bar{v}\delta \cdot d\bar{v} \tag{7.6}$$

and the spectrum computed on transforming this interferogram would be $\tfrac{1}{2} I(\bar{v})[1 - T(\bar{v})]$.

It can be seen that intense bands in a spectrum measured in this fashion can be strong for two reasons, either the intensity of the source, $I(\bar{v})$, at that frequency is large or the absorption of the radiation, $1 - T(\bar{v})$, is high.

Figure 7.1 shows a low-resolution difference spectrum of a polystyrene film, together with the transmittance spectrum and the single-beam background for this measurement. The three strong bands in the difference spectrum at about 1600, 1500, and 1450 cm^{-1} all have a fairly high absorption, but in addition they are located in a region of the spectrum where the single-beam energy of the source is large. In the transmittance spectrum, the overtone and combination bands between 1700 and 2000 cm^{-1} are much weaker than the two intense C–H bending bands at 700 and 760 cm^{-1}. However, since the single-beam energy was much greater between 1700 and 2000 cm^{-1} than it

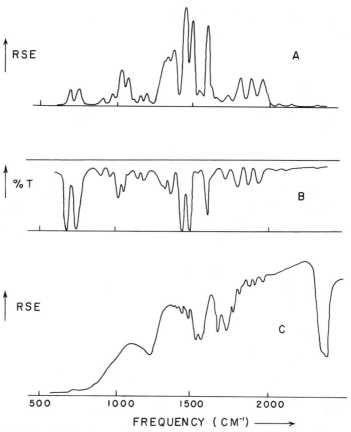

Fig. 7.1. (a) The absorption spectrum of polystyrene measured by optical subtraction; (b) the absorption spectrum of polystyrene plotted linear in transmittance, measured under approximately the same resolution as (a); (c) the background emission spectrum of the source used in the measurement of (a).

was between 650 and 800 cm^{-1}, the relatively weak overtone and combination bands in the difference spectrum appear to be more intense than the very strong C–H bending bands. Thus, some care has to be taken in the interpretation of spectra measured using the optical subtraction technique.

The real importance of this method is the tremendous reduction in the amplitude of the interferogram at zero retardation. Let us consider the same example used earlier in this chapter to illustrate the dynamic range problem with weak absorption bands when a single input beam and output beam are measured but this time employing dual-beam FTS to reduce the dynamic range. If the only absorption band in the spectrum has a maximum absorption of 0.1% (with $\sigma_s = 6$ cm^{-1}, $\sigma_b = 3000$ cm^{-1}) the dynamic range required for its characterization using dual-beam FTS is 2^{19} less than if conventional single-beam techniques were applied (assuming that a perfect null can be obtained when no sample is present in either beam). The problem now becomes one of detection rather than one of dynamic range. Whereas with a single input beam sensitive detectors cannot be used because of dynamic range considerations, with dual-beam input sensitive detectors can be easily used. In this way extremely small absorption bands should be able to be measured.

Low [3] has described an experimental arrangement designed for the measurement of weak absorption bands of GC effluents; his system is shown in Fig. 7.2. The ratio of the amplitude of the interferogram at zero retardation using a single input beam to the amplitude using the dual-beam input from both sources has been termed the *nulling ratio* by Low. If the signals from each source are perfectly matched, the nulling ratio should be infinite. Using the optical arrangement shown in Fig. 7.2, Low was able to achieve a nulling ratio of 30:1 with the two gas-cells cold and 20:1 with them hot. In a later system using the same principle, Low and Mark [4] achieved a nulling ratio of 100:1.

Part of the reason why a better nulling ratio could not be achieved with either system might be assigned to the fact that physically different sources were used for each input beam. Bar-Lev [5] described a system using one source and two detectors, so that if the two detectors were perfectly matched a high nulling ratio should have been attained. With this arrangement Bar-Lev achieved an experimental nulling-ratio of 60:1.

In an attempt to achieve a higher nulling ratio, Griffiths and Lephardt [6] designed an arrangement using a single source and detector. Like Low's [3] experiment, this system was designed for the purpose of measuring the infrared spectra of GC peaks. Using this system, which is shown diagrammatically in Fig. 7.3, Griffiths and Lephardt also attained a nulling ratio of about 100:1, even when the gas-cells were heated, corresponding to a reduction of almost 2^7 in dynamic range which was more than sufficient for the

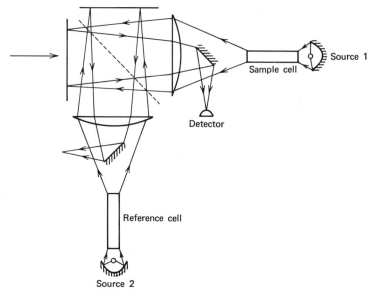

Fig. 7.2. Optical layout of the dual-beam system used by Low [3] for the measurement of the spectra of GC peaks. (Reproduced from [7] by permission of Marcel Dekker Inc.; copyright © 1972.)

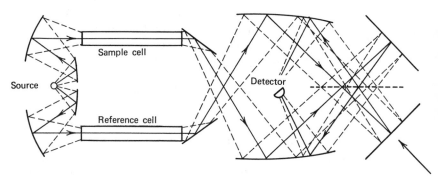

Fig. 7.3. Optical layout of the dual-beam system used by Griffiths and Lephardt [6] for the measurement of the spectra of GC peaks. In this system, only one source was used. (Reproduced from [7] by permission of Marcel Dekker Inc.; copyright © 1972.)

purpose for which the experiment was designed. This design has an advantage over the previously mentioned designs of Bar-Lev and Low in that no pick-off mirror is present in the beam so that theoretically the total energy input to the interferometer should reach the detector. This arrangement could also be used for higher resolution measurements than Bar-Lev's or Low's arrangements, since both of the latter systems require that the divergence angle of the beam passing through the interferometer is rather large. For very high resolution measurements using the dual-beam technique[7] the system of Burroughs and Chamberlain [8] described in Section II is felt to present the greatest potential.

Both Low and Mark [4] and Griffiths and Lephardt [6] found that the residual spectrum after optical subtraction did not have the same profile as the single-beam background. Figure 7.4 shows the transform of an interferogram measured by Griffiths and Lephardt at a nulling ratio of about

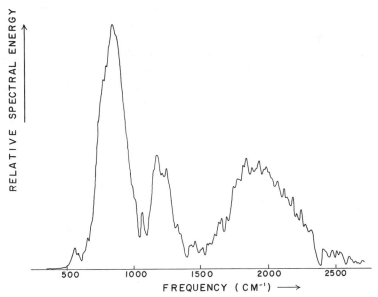

Fig. 7.4. Transform of an interferogram measured using the optical subtraction method. The two bands centered at about 850 and 125 cm^{-1} may be caused by species on the surface of the beamsplitter, whereas the energy at higher frequency is probably a result of imperfect matching of the two beams.

80:1. The broad signal centered at about 1800 cm^{-1} appears to be caused by imperfect matching of the two beams, since its amplitude increased as the degree of matching was reduced. However the magnitude of the two low frequency maxima stayed essentially constant as the system was brought in and out of alignment. It is felt that these features could possibly be caused by oxides of germanium on the upper surface of the beamsplitter (adjacent to the compensator plate). The magnitude of these two features is considerably greater than the noise level, and it was found that by measuring a background interferogram with both cells empty and subtracting this digitally from the interferogram measured with the sample in one cell allowed a perfectly flat background to be achieved, limited only by detector noise. It should again be stressed that the optical subtraction method is only useful when the signal-to-noise ratio of a spectrum is limited by the dynamic range of the digitizer. Most rapid-scanning Fourier transform spectrometers designed for absorption measurements in the mid infrared use a TGS detector whose specific detectivity D^* is usually sufficiently small that the signal-to-noise ratio of the unattenuated interferogram does not exceed the dynamic range of a 15-bit ADC. Recently (late 1974), Spectrotherm Corp. [9] has introduced a rapid-scanning dual-beam interferometer for the specific purpose of measuring the infrared spectra of gas chromatographic peaks on-line to much lower detection limits than was possible with previously available equipment. The key component of this instrument is a very sensitive mercury cadmium telluride detector whose D^* is such that the optical subtraction technique is mandatory to observe bands absorbing 0.01% of the incident radiation. The optical layout of this spectrometer is shown in Fig. 7.5 and it shows several improvements over previously described systems. First, no pick-off mirrors are needed so that essentially the full energy from the source

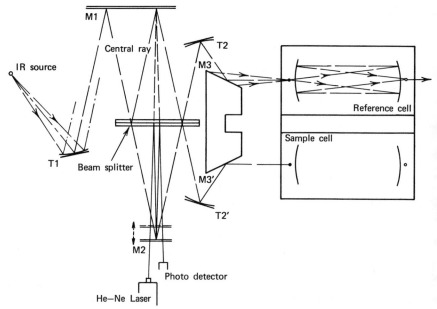

Fig. 7.5. Optical layout of the Spectrotherm dual-beam Fourier transform spectrometer configured for the on-line measurement of GC peaks. (a) showing how the two exit beams pass through the sample and reference cells; (b) showing how these two beams are combined and passed onto the detector. (Courtesy of Spectrotherm Corporation.)

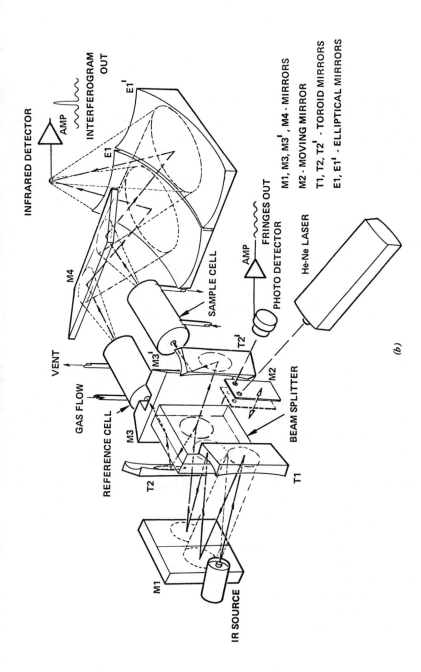

INFRARED DETECTOR

AMP

INTERFEROGRAM OUT

E1'

E1

M4

VENT

GAS FLOW

REFERENCE CELL

M3

M3'

SAMPLE CELL

T2

T2'

M2

BEAM SPLITTER

T1

FRINGES OUT

AMP

PHOTO DETECTOR

He-Ne LASER

M1, M3, M3', M4 - MIRRORS

M2 - MOVING MIRROR

T1, T2, T2' - TOROID MIRRORS

E1, E1' - ELLIPTICAL MIRRORS

M1

IR SOURCE

(b)

179

reaches the detector. In addition, the gas cell has been designed so that the throughput is matched to the rest of the spectrometer. Without the use of dual-beam Fourier transform spectroscopy it is apparent that the development of this important instrument would not have been possible.

Low and Mark [4] applied the optical subtraction technique to the measurement of surface coatings and films on metallic reflectors. They found that it was relatively easy to achieve a high nulling ratio if the substrates are front surface mirrors, and that there is little difficulty in obtaining a usable optically subtracted spectrum if the sample substance is a liquid which spreads evenly over the mirror surface on which it is placed. On the other hand they found that solid samples cause great difficulty if a smooth film is not formed and that in extreme cases the optical subtraction technique becomes of little value. It may be noted that in all of these extreme cases, strong absorption bands were seen in the single-beam spectra so that the dual-beam technique is unnecessary. It seems very probable to this author that dual-beam Fourier transform spectroscopy could become of great importance in the study of monomolecular layers of species absorbed on specularly reflecting metal surfaces; for this type of sample the absorbance of the bands is usually very low and often less than 0.1%.

II. REDUCTION OF NOISE

A further practical benefit of dual-beam Fourier transform spectroscopy is the reduction of the noise level in the spectrum. For example, Burroughs and Chamberlain [8] have studied very far-infrared atmospheric absorption spectra using the sun as the source. Deviations from the true interferogram may arise during this measurement not only because of detector noise (which is additive in nature), but also because of such factors as passing clouds, poor following of the source, fluctuations in the atmospheric transparency or fluctuations of the refractive index of the atmosphere. All these factors affect the amplitude of both the a.c. and d.c. components of the interferogram, $I'(\delta)$, so that even the "d.c." component now fluctuates in time. If the frequency bandwidth of these fluctuations overlaps the bandwidth of the a.c. components of the interferogram, the result will be an increase in the noise level of the spectrum. Mathematically, the effect of these factors may be represented by *multiplying* both the a.c. and d.c. components of $I'(\delta)$ by a randomly varying function, $E(\delta)$; hence noise of this type of origin is usually called *multiplicative noise*. The general form of an interferogram exhibiting both multiplicative and additive noise is:

$$J(\delta) = I'(\delta) \cdot E(\delta) + D(\delta) \qquad (7.7)$$

where $D(\delta)$ is the additive (detector) noise. Burroughs and Chamberlain [8] have shown that the effect of multiplicative noise can be greatly reduced using dual-beam FTS and they experimentally confirmed that additive noise is also reduced when this technique is used.

For a single input beam, let the portion that is transmitted through the interferometer be again referred to with the subscript t, while that which is reflected back to the source will be referred to using the subscript r. Abbreviating the terms in Eqs. 7.1 through 7.4, let I_t be the constant component of the signal transmitted through the interferometer, I_r be that reflected back to the source, and $F(\delta)$ be the a.c. component of the interferogram. For a nonabsorbing beamsplitter $F(\delta)$ will be the same for both the reflected and the transmitted beams. Thus:

$$I_t'(\delta) = I_t + F(\delta) \tag{7.8}$$

and

$$I_r'(\delta) = I_r - F(\delta). \tag{7.9}$$

If additive noise and multiplicative noise are both present, the interferogram measured at the detector for the transmitted beam is

$$J_t(\delta) = I_t E(\delta) + F(\delta)E(\delta) + D_t(\delta) \tag{7.10}$$

while for the reflected beam it is

$$J_r(\delta) = I_r E(\delta) - F(\delta)E(\delta) + D_r(\delta). \tag{7.11}$$

The difference between $J_t(\delta)$ and $J_r(\delta)$ augments the a.c. interferogram but reduces the value of the constant terms:

$$
\begin{aligned}
J_d(\delta) &= J_t(\delta) - J_r(\delta) \\
&= (I_t - I_r)E(\delta) + 2\,F(\delta)E(\delta) + D_t(\delta) - D_r(\delta)
\end{aligned}
\tag{7.12}
$$

For an ideal beamsplitter, $I_t = I_r$, so that the first term of Eq. 7.12 disappears and no modulation of the d.c. component of the interferogram due to multiplicative noise is seen. Even in the case of the Mylar beamsplitter, which is far from ideal, this background term becomes much smaller than if the dual-beam technique were not used.

If the effect of multiplicative noise is negligible compared to the detector noise, the difference signal becomes equal to

$$J_d(\delta) = 2\,F(\delta)E(\delta) + D_t(\delta) - D_r(\delta) \tag{7.13}$$

If the noise-equivalent powers of both detectors are equal, the noise level in the interferogram, $[D_t(\delta) - D_r(\delta)]$, will be greater than either $D_t(\delta)$ or $D_r(\delta)$ by a factor of $\sqrt{2}$, while the amplitude the a.c. component of the interferogram has doubled. Thus the signal-to-noise ratio in the spectrum is increased by a factor of $\sqrt{2}$.

Figure 7.6 shows some measured traces of Burroughs and Chamberlain

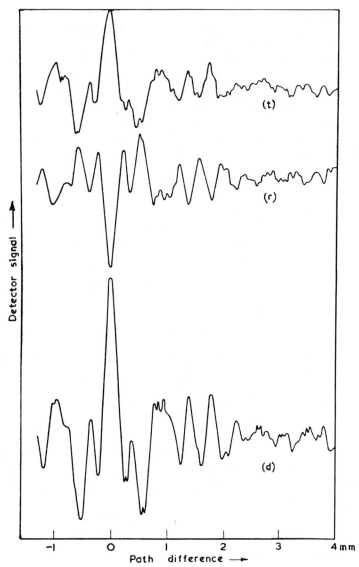

Fig. 7.6. Interferograms measured by Burroughs and Chamberlain from both output beams of their interferometer. (*t*) is the "transmitted" interferogram, (*r*) is the "reflected" interferogram, and (*d*) is the difference between them. (Reproduced from [8] by permission of the author and Pergamon Press; copyright © 1971.)

182

[8] for the transmitted (t), reflected (r), and difference (d) interferograms. The noise level in the t and r traces is approximately the same and the signal levels are also approximately equal, although the r trace is inverted with respect to the t trace. However, the signal level of the d trace is twice as great as either of the other two. Burroughs and Chamberlain found slight differences between spectra from the transmitted and reflected interferograms which are probably related to absorption in the beamsplitter.

If any radiation is absorbed by the beamsplitter material, the effect will be seen to a greater extent in the transmitted than the reflected interferogram. For example, in Fig. 7.6 the first minimum on either side of the zero retardation point has a smaller amplitude in the t interferogram. This minimum is caused largely by radiation from the second hoop of the beamsplitter efficiency curve, and since the absorbance of Mylar increases with frequency and a thick (190 μm) film was used for this measurement, the energy transmitted through the interferometer would indeed be expected to be less for the higher frequencies present in the spectrum.

Some information on the frequency distribution of the noise is provided by the *sum* of the interferograms $J_t(\delta)$ and $J_r(\delta)$

$$J_s(\delta) = (I_t + I_r)E(\delta) + D_t(\delta) + D_r(\delta). \qquad (7.14)$$

The frequency dependence of the detector noise, $D(\delta)$, is easily measured by covering the entrance aperture of the detector and measuring an "interferogram" with the same sampling frequency that is used in the spectral measurements. The transform of this signal shows the variation of the detector noise with frequency. Fourier transform of the "sum" interferogram measured with the movable mirror of the interferometer held stationary, again using the same sampling frequency, gives the sum of the multiplicative and additive noise as a function of frequency. These noise "spectra" should be determined by averaging as many of the "sum" interferograms as possible.

Comparison of the transforms of the signals measured with the entrance aperture open and shut permits one to determine if multiplicative or additive noise is the limiting factor. When multiplicative noise is present, the noise power is often greatest at low frequency, since the fluctuations in source intensity usually occur slowly. If this is the case, increasing the scan speed of the interferometer can shift the range of modulation frequencies being studied ($2V\bar{\nu}_{max}$ to $2V\bar{\nu}_{min}$, see Eq. 1.10) away from the frequency range of the source fluctuations.

Burroughs and Chamberlain [8] used an interferometer with retroreflecting mirrors to prevent the beam from returning directly to the source (Fig. 7.7). The use of retroreflecting mirrors is probably the optimum method for measuring interferograms if either two sources or two detectors are being employed. With this system, no pick-off mirrors such as those shown in

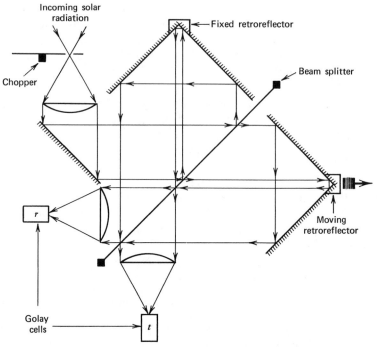

Fig. 7.7. Interferometer with retroreflectors used by Burroughs and Chamberlain, enabling the reflected output beam to be easily spatially separated from the input beam. (Reproduced from [8] by permission of the author and Pergamon Press; copyright © 1971.)

Fig. 7.2 are needed so that the energy at the detector is increased. The system shown in Fig. 7.3 is difficult to align and has the disadvantage that the beam is not incident at the beamsplitter at 45°, which can occasionally cause problems. The only disadvantage to the use of retroreflecting mirrors in the interferometer is that a wider aperture for the instrument is required, and the complete area of the beamsplitter is not used, but these factors are quite minor compared to the advantages of this design.

III. THE POLARIZATION INTERFEROMETER

A rather different type of interferometer where both of the output beams are measured has been described by Martin [10, 11]. In this instrument, the beamsplitter is a *polarizer* producing transmitted and reflected beams which are orthogonally plane-polarized. The two output beams are differentiated not by their direction of propagation, as is the case for a conventional Michelson interferometer, but by their plane of polarization.

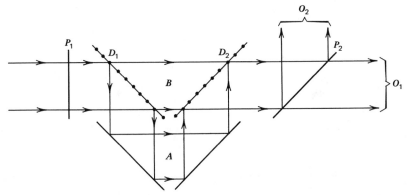

Fig. 7.8. Diagram illustrating the nature of the polarization interferometer. (Reproduced from [11] by permission of the author and the D. Reidel Publishing Co.; copyright ⓒ 1972.)

Figure 7.8 shows the essential features of the method. A collimated beam is plane-polarized at P_1 in the plane at 45° to the normal to this page. It is then divided into beam A, polarized with its electric vector normal to the paper, and beam B, polarized at 90° to A, by a flat wire-grid polarizer D_1. A and B are recombined at wire-grid D_2 and the combined beam finally passes through polarizer P_2 which has its axis parallel to that of P_1 (or at 90° to that direction). With a monochromatic source the beam is elliptically polarized with an ellipticity varying periodically with increasing path-difference between beam A and beam B. After P_2 the beam is plane-polarized with an amplitude which varies with path-difference in the same way as in a conventional Michelson interferometer, that is,

$$I_p'(\delta) = 0.5\, I(\bar{v}) \left\{ 1 + \cos 2\pi\bar{v}\delta \right\} \qquad (7.15)$$

$$I_t'(\delta) = 0.5\, I(\bar{v}) \left\{ 1 - \cos 2\pi\bar{v}\delta \right\} \qquad (7.16)$$

The case I_p' is for parallel P_1 and P_2, and I_t' is for crossed P_1 and P_2.

Since these two interferograms are 180° out-of-phase, it is possible to eliminate the d.c. level of the interferogram by alternating the orientation of P_1 instead of chopping in the usual fashion. Alternatively, the signal from the detector measuring $I_t'(\delta)$ could be subtracted from the signal from the other detector. In either case the output for a monochromatic source is:

$$I(\delta) = I_p'(\delta) - I_t'(\delta) = I(\bar{v}) \cos 2\pi\bar{v}\delta \qquad (7.17)$$

which does not contain a d.c. component.

Figure 7.9 illustrates the system used in practice by Martin. Corner reflectors are used as rotators of the plane of polarization to give a configuration close to that of a conventional Michelson interferometer, and in fact Martin's

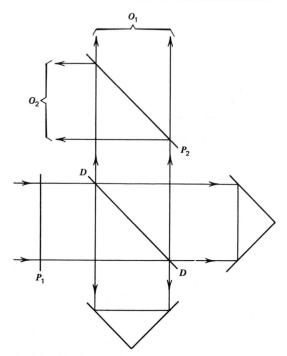

Fig. 7.9. The layout of the polarization interferometer used by Martin for far-infrared spectroscopy; the interferometer is a modified Beckman–RIIC FS–720. (Reproduced from [11] by permission of the author and the D. Reidel Publishing Co.; copyright © 1972.)

measurements have been made on a modified Beckman-RIIC FS–720 interferometer. The polarizer P_1 is placed close to the source and is rotated, alternating the orientation of P_1 between the parallel and perpendicular settings with respect to P_2. The depth of modulation is as great as that for a thin-film beamsplitter working at maximum efficiency.

Martin has made the point that the efficiency of far-infrared polarization interferometers remains high over longer frequency ranges than that of the conventional Michelson interferometer since wire-grid polarizers have good efficiencies at very low frequency and maintain this performance down to wavelengths comparable to the spacing of the grid. Thus, with a polarization interferometer it is currently possible to work from close to zero frequency up to 100 cm^{-1} with nearly constant efficiency. For this reason Martin has claimed that this type of instrument should displace the lamellar grating interferometer for very low frequency spectroscopy.

REFERENCES

1. C. H. Perry, R. Geick, and E. F. Young, *Appl. Optics*, **5**, 1171 (1966).
2. F. Block and P. R. Griffiths, *Appl. Spectrosc.*, **27**, 431 (1973).
3. M. J. D. Low, *Anal. Letters*, **1**, 819 (1968).
4. M. J. D. Low and H. Mark, *J. Paint Technol.*, **43**(No. 553), 31 (1971).
5. H. Bar-Lev, *Infrared Phys.*, **7**, 93 (1967).
6. P. R. Griffiths and J. O. Lephardt, paper presented at the Pittsburgh Conf. of Anal. Chem. and Appl. Spectrosc. (1969).
7. P. R. Griffiths, C. T. Foskett and R. Curbelo, *Appl. Spectrosc. Revs.*, **6**, 31 (1972).
8. W. J. Burroughs and J. Chamberlain, *Infrared Phys.*, **11**, 1 (1971).
9. Spectrotherm Corporation, 3040 Olcott Street, Santa Clara, Calif., 95051.
10. D. E. Martin and E. Puplett, *Infrared Phys.*, **10**, 105 (1969).
11. D. E. Martin, *Infrared Detection Techniques for Space Research*, V. Manno and J. Ring, Eds., D. Reidel Publishing Co., Dordrecht, Holland, 1972.

SIGNAL-TO NOISE RATIO IN FTS

I. DETECTOR NOISE

The most basic and unavoidable of all types of noise in a spectrum measured using Fourier transform techniques is detector noise [1], and every system should preferably be designed so that detector noise is greater than noise arising from any other source.

The sensitivity of infrared detectors is commonly expressed in terms of the *noise equivalent power* (NEP) of the detector, which is the ratio of the detector noise voltage, V_n in volts/Hz$^{\frac{1}{2}}$, to the voltage responsivity, R_v in volts per watt, that is,

$$\text{NEP} = \frac{V_n}{R_v}. \tag{8.1}$$

The NEP is dependent on the area of the detector, and a function which is usually called the *specific detectivity*, D^*, has been defined which is to a good approximation independent of detector area, A_D, for most materials:

$$D^* = \frac{A_D^{\frac{1}{2}}}{\text{NEP}}. \tag{8.2}$$

The *noise power* observed in a measurement time, t sec, is given by

$$N = \frac{\text{NEP}}{t^{\frac{1}{2}}}. \tag{8.3}$$

To determine the signal-to-noise ratio obtainable from any source, we must not only know the noise power but also the power of the signal. The spectral energy density from a black-body source, $u_{\bar{v}}(T)$ is given as a function of temperature, T, by the familiar equation first derived by Planck:

$$u_{\bar{v}}(T) = \frac{C_1 \bar{v}^3}{\exp\left(-C_2 \bar{v}/T\right) - 1} \text{ W/sr-cm}^2\text{-cm}^{-1} \tag{8.4}$$

where C_1 and C_2 are the first and second radiation constants, respectively. The power received at a detector through any optical system is determined by the *throughput* of that system, that is, the product of the area of the beam and its solid angle at any focus. For an optimally designed Fourier transform spectrometer, the throughput is determined by the area of the mirrors of the interferometer and the maximum allowed solid angle, which may be calculated using Eq. 1.32. The power received at a detector through an interferometer having a throughput θ, a resolution $\Delta\bar{v}$, and an efficiency η, in unit frequency interval, is given by

$$S' = u_{\bar{v}}(T) \cdot \theta \cdot \eta \cdot \Delta\bar{v}. \tag{8.5}$$

Thus the signal-to-noise ratio of a spectrum measured using a Michelson interferometer is given by

$$\frac{S'}{N} = \frac{u_{\bar{v}}(T) \cdot \theta \cdot \Delta\bar{v} \cdot t^{\frac{1}{2}} \cdot \eta}{\text{NEP}}. \tag{8.6}$$

Since the resolution, $\Delta\bar{v}$, is given by Eq. 1.14 as the reciprocal of the maximum retardation, Δ_{max}, the signal-to-noise ratio may also be expressed as

$$\frac{S'}{N} = \frac{u_{\bar{v}}(T) \cdot \theta \cdot \eta \cdot t^{\frac{1}{2}}}{\text{NEP} \cdot \Delta_{max}}. \tag{8.7}$$

Another useful parameter, sometimes called the *noise equivalent spectral radiance* (NESR), has been defined giving the spectral energy density for which the signal-to-noise ratio for a given interferometer-detector combination is unity. From Eq. 8.7, it is apparent that

$$\text{NESR} = \text{NEP} \cdot \frac{\Delta_{max}}{\theta \cdot \eta \cdot t^{\frac{1}{2}}} \tag{8.8}$$

or

$$\text{NESR} = \frac{\Delta_{max} \cdot A_D^{\frac{1}{2}}}{D^* \cdot \theta \cdot \eta \cdot t^{\frac{1}{2}}} \tag{8.9}$$

where Δ_{max} is the maximum retardation in cm;
$\quad$ θ is the limiting optical throughput in cm^2-steradian;
$\quad$ η is the overall efficiency of the system (including
$\quad\quad$ beamsplitter efficiency, transmission losses, etc.);
$\quad$ t is the total measurement time in seconds;
$\quad$ A_D is the detector area in cm^2.

One of the most important parameters in this equation is the optical throughput, θ. The throughput can, in practice, be limited either by the maximum allowed angle of the beam, as given by Eq. 1.32, or by the physical constraints of the optics, especially the detector size or the $f/$ number of the condensing optics in front of the detector. The latter effect is commonly encountered for low resolution measurements where a large throughput is allowed for the interferometer. In this case, the throughput is said to be detector-size limited, and is given the symbol, θ_D. θ_D is equal to the product of the solid angle of the beam being focused on the detector Ω_D steradians and the detector area A_D, that is,

$$\theta_D = A_D \Omega_D \text{ sr-cm}^2 \qquad (8.10)$$

When the solid angle of the beam through the interferometer is determined by the maximum frequency in the spectrum, $\overline{\nu}_{max}$, and the desired resolution, $\Delta\overline{\nu}$, the solid angle of the beam through the interferometer θ_I is given by

$$\Omega_I = 2\pi \left(\frac{\Delta\overline{\nu}}{\overline{\nu}_{max}}\right) \text{ sr} \qquad (8.11)$$

so that θ_I is given by

$$\theta_I = \frac{2\pi A_M (\Delta\overline{\nu})}{\overline{\nu}_{max}} \text{ sr-cm}^2 \qquad (8.12)$$

where A_M is the area of the mirrors in the interferometer.

To determine whether θ_I or θ_D is the parameter that should be used in Eq. 8.9 for a given Fourier transform spectrometer, the relative magnitudes of A_D, A_M, $\Delta\overline{\nu}$, $\overline{\nu}_{max}$, and Ω_D should be compared. This is quite easily done, and the procedure will be illustrated by using the parameters for a typical Fourier transform spectrometer:

$A_D = 0.04 \text{ cm}^2$ (2 mm square detector)
$A_M = 17.5 \text{ cm}^2$ (2 in. diameter mirrors)
$\Delta\overline{\nu} = 1 \text{ cm}^{-1}$ (arbitrarily)
$\overline{\nu}_{max} = 4000 \text{ cm}^{-1}$ (for mid-IR spectroscopy)
$\Omega_D = 1.5 \text{ sr}$ ($f/1$ condensing optics)

Thus $\theta_D = 0.04 \times 1.5 = 0.06 \text{ sr-cm}^2$

and $\theta_I = \dfrac{2 \times 17.5}{4000} = 0.03 \text{ sr-cm}^2$

Since θ_I is less than θ_D, the throughput is limited by the maximum angle of the beam through the interferometer for 1 cm^{-1} resolution.

If a resolution of 2 cm^{-1} is desired and θ_D remains constant while θ_I is doubled, θ_I becomes equal to θ_D. At this point the optics are said to be *throughput matched*. This condition represents the best operating state for any Fourier transform spectrometer since the detector is exactly filled by the image of the source while the beam passes through the interferometer with the maximum allowed solid angle. Any further increase in the throughput would result in the image of the source at the detector being larger than A_D so that the throughput would remain constant at the value given by θ_D.

The Digilab [2] series of Fourier transform spectrometers nicely illustrate the ramifications of this theoretical discussion on instrumental design. In the FTS–14 spectrometer the throughput is limited by the detector size for any resolution lower than 2 cm^{-1}, so that the throughput is always constant on this instrument at low resolution. The spectrometer is throughput matched at 2 cm^{-1} resolution; consequently, the beam has to be stopped down to reduce the solid angle through the interferometer wherever measurements are taken over the complete mid-infrared spectrum at any resolution higher than 2 cm^{-1}.

In this instrument, only one aperture stop is used which optimizes the throughput for a resolution of 0.5 cm^{-1}. Thus, correct values for θ_I can only be obtained for resolution settings of 2 cm^{-1} and 0.5 cm^{-1}. For this reason, in the discussion of "trading rules" in the following section, the variation of S/N with resolution and measurement time under conditions when θ_I is optimized is illustrated with spectra measured at 2 cm^{-1} and 0.5 cm^{-1} resolution. For higher resolution interferometers such as the Digilab FTS–20, the EOCOM FMS 7201, and the Idealab IF–6, each of which can achieve a resolution of at least 0.1 cm^{-1}, many interchangeable aperture stops are provided to optimize the throughput for any combination of $\Delta \bar{v}$ and $\bar{v}_{max}$. The way in which the signal-to-noise ratio varies with resolution is strongly dependent on whether the interferometer is being used under the constant throughput (θ_D) or variable throughput (θ_I) condition, and this effect will be discussed in some detail in the next section.

The noise equivalent spectral radiance is easily calculated for a typical mid-infrared Fourier transform spectrometer using the parameters listed above together with an estimate for the overall efficiency of the system ($\eta = 0.1$) and a knowledge of the detectivity of the TGS detector ($D^* = 10^8$ cm Hz$^{\frac{1}{2}}$ W^{-1}). For a measurement time of 15 minutes ($t = 900$ sec) the NESR is given by

$$\text{NESR} = \frac{1 \times 0.04}{10 \times 0.03 \times 30 \times 0.1}$$
$$= 2.2 \times 10^{-8} \text{ W/sr-cm}^2\text{-cm}^{-1}$$

To calculate the S/N at any frequency for a spectrum measured under these conditions, the radiance of the source, $u_{\bar{v}}(T)$, must be known. For a 150°C black-body, N is equal to 4×10^{-6} W/sr-cm^2-cm^{-1} at 1000 cm^{-1}. Thus the S/N at 1000 cm^{-1} measured under these conditions should be equal to

$$\text{S/N} = \frac{u_{1000}(423)}{\text{NESR}} = \frac{4 \times 10^{-6}}{2.2 \times 10^{-8}} = 180.$$

In practice, a S/N of about 200 is found for measurements of black-body sources at 150°C, so that it can be seen that this method of estimating the sensitivity of any measurement gives quick, valid results.

II. "TRADING RULES" IN FTS

The quantitative relationships between S/N, resolution, and measurement time are commonly referred to as "trading rules" [3]; the trading rules for Fourier transform spectroscopy are quite easily worked out using Eq. 8.9.

For any spectrometer, conventional or interferometric, the S/N of a spectrum measured at a given resolution is proportional to the square root of the measurement time. The dependence of the NESR on $t^{-\frac{1}{2}}$ shown is Eq. 8.9 demonstrates that this is true for FTS. To change the time taken to measure an interferogram, one can either change the scan speed of the moving mirror of the interferometer or repetitively scan the interferogram at a constant mirror velocity and average either the interferograms or the spectra found by computing the Fourier transform of each interferogram.

When the scan speed of an interferometer is changed, care has to be taken to ensure that the desired effect on the spectrum is achieved. Consider the case of a slow-scanning interferometer where a chopper is used to modulate the beam at a certain frequency. If the scan speed is reduced by a factor of two and the time constant of the filter is left unchanged, no change would be noticed in the S/N of the spectrum in spite of the change in measurement time. The only change that has been made is a reduction of the observation efficiency. If the scan-speed is increased by a factor of two without changing the time constant of the filter, the shorter wavelengths in the spectrum will be modulated at a sufficiently high frequency that they will be attenuated by the filter. The time constant on the filter must always be sufficiently long that high noise frequencies are not present in the interferogram or else a rather high sampling frequency would be required if noise is not going to be folded back into the spectrum. Thus, any time the scan-speed of the interferometer mirror is changed, the time constant of the filter must be changed by a corresponding amount.

The situation is rather more simple for rapid-scanning interferometers; here the measurement time is increased by increasing the number of scans to be signal-averaged and no change of the filter pass-band is necessary. To increase the S/N by a factor of two, the number of scans to be signal-averaged is increased by a factor of four.

Consider now the effect of increasing the resolution of a spectrum measured with a given number of scans (for a rapid-scan system) or at a constant scan-speed for a slow-scan interferometer. If the retardation, Δ_{max}, is increased by a factor of two, the measurement time, t, is also increased by a factor of two. Since the NESR is proportional to $\Delta_{max}/t^{\frac{1}{2}}$, the noise is increased by a factor of $\sqrt{2}$ in this case. The effect of the difference in NESR on the S/N of the spectrum must be examined for the constant throughput and variable throughput case separately.

A.   THE CONSTANT THROUGHPUT CASE

Under conditions of constant optical throughput, the signal at the detector does not change and therefore the S/N in a spectrum measured by doubling the retardation is degraded by a factor of $\sqrt{2}$ while the measurement time is doubled. To obtain a S/N identical to that of the lower resolution spectrum, the measurement time must be further doubled, making the total increase in measurement time a factor of four.

This rule has been verified [4] using an FTS–14 spectrometer. Spectra were measured at 2 cm^{-1} and 4 cm^{-1} resolution at constant throughput and it was found that twice the number of scans were needed at 2 cm^{-1} resolution to achieve the same noise level as a 4 cm^{-1} resolution spectrum (Fig. 8.1).

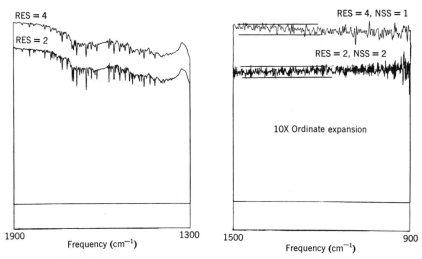

Fig. 8.1.   Constant throughput case: (a) single-beam spectra of water vapor measured at resolutions of 4 cm^{-1} (RES = 4) and 2 cm^{-1} (RES = 2). (b) ratio of consecutive single-beam spectra (100% line) under 10x scale expansion: the time for a single scan at RES = 2 is 1.6 sec, and for RES = 4 it is 0.8 sec, NSS refers to the number of scans signal averaged, so that the measurement time for the upper trace is one-fourth that of the lower trace, and the noise levels in each spectrum are identical. (Reproduced from [4] by permission of the American Chemical Society; copyright © 1972.)

B.   THE VARIABLE THROUGHPUT CASE

Under the variable throughput criterion the same rule governing NESR applies, but here the signal at the detector is reduced when the resolution is increased. If the resolution is increased by a factor of two (numerically halved) the solid angle of the beam through the interferometer (and hence the signal at the detector) must be reduced by a factor of two as shown by Eq. 1.32. Therefore, the measurement time must be further increased by a factor of 2^2 over the constant throughput case. Therefore, to obtain a S/N identical to that of the lower resolution spectrum, the measurement time has to be increased by a factor of sixteen in total. Figure 8.2 shows that this rule is

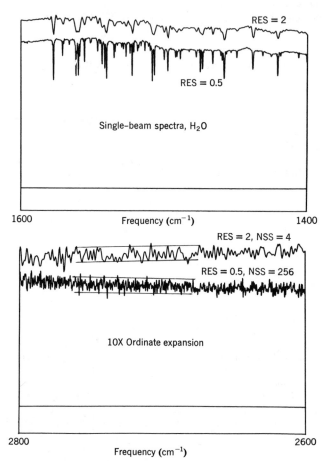

Fig. 8.2. Variable throughput case: (a) single-beam spectra measured at RES = 2 (upper trace) and RES = 0.5. The time for a single scan at RES = 0.5 is 6.4 sec; (b) ratio of consecutive scans under 10x scale expansion; peak-to-peak noise levels are identical for 4 scans at 2 cm^{-1} resolution and 256 scans at 0.5 cm^{-1} resolution. (Reproduced from [4] by permission of the American Chemical Society; copyright © 1972.)

valid for the case of spectra measured when the throughput is limited by the maximum allowed angle of the beam passing through the interferometer.

This relationship between S/N, measurement time and resolution is exactly the same as the corresponding rule for grating spectrometers, which shows that the advantage of a given interferometer over a given dispersion spectrometer should always be constant and should not increase with resolution. On the other hand, for the case when the optical throughput of the interferometer remains constant the interferometer shows its greatest advantage over the dispersion spectrometer at the highest resolution. Whereas the sensitivity of a dispersion spectrometer will always decrease as the resolution is increased, this is not always the case with Fourier transform spectrometers used at constant throughput as can be verified by the following example.

Consider the case of a *weak* isolated narrow absorption band whose percent absorption is $A\%$ when measured at 4 cm^{-1} resolution and approximately $2A\%$ when measured at 2 cm^{-1} resolution, and let us examine how the absorption intensity-to-noise ratio (I/N) is affected if the spectrum is measured on a dispersive and a Fourier transform spectrometer with variable throughput. If the noise level on the spectrum measured with a grating spectrometer at 4 cm^{-1} resolution is $x\%$, I/N $= A/x$. For a spectrum measured over the same spectral range in the same measurement time at 2 cm^{-1} resolution, the noise level is $4x\%$, so that I/N $= 2A/4x = \frac{1}{2}A/x$. Thus there is a loss of sensitivity on increasing the resolution, and since the trading rules are the same for a dispersion spectrometer and a Fourier transform spectrometer used at variable throughput, precisely the same rules governing this loss of sensitivity are found for an interferometer used at variable throughput and a grating spectrometer.

If the same band is studied with a constant throughput interferometer, there is a different relationship. Let the noise level be $x\%$ when the spectrum is measured with N scans at 4 cm^{-1} resolution. If the same spectrum is measured at 2 cm^{-1} resolution with N scans, the noise level would be $\sqrt{2}x\%$, so that I/N $= 2A/\sqrt{2}x = \sqrt{2}A/x$. These spectra were measured with equal numbers of scans, so that if the spectra are to be measured with equal measurement times, $2N$ scans would be required of the lower resolution measurement, there by reducing the noise level to $x/\sqrt{2}\%$. The value of I/N under these conditions is $\sqrt{2}A/x$, which demonstrates exactly the same sensitivity as the higher resolution spectrum run with the same measurement time. In view of the slight improvement in the efficiency of the duty cycle for a rapid-scanning interferometer as the resolution is increased, there may actually be a slight increase in sensitivity on increasing the resolution of a constant throughput Fourier transform spectrometer for a given measurement time.

If the band being measured is *broad* relative to the resolution, no such improvement is noted since the absorption intensity does not increase as the resolution is increased. In this case the maximum sensitivity is always found with the lowest resolution measurements. Pickett and Strauss [5] showed this to be the case for far-infrared spectroscopy, and stated that in order to obtain the best sensitivity, spectra should be run with a resolution equal to the width of the band under investigation. In the opinion of this author, spectra should be run at higher resolution than this, since noise in an interferometrically measured spectrum has the same period as the resolution at which the spectrum is measured. Thus, it becomes very difficult to distinguish between weak real spectral features and noise when the spectrum is run at the resolution suggested by Pickett and Strauss. In practice, it has been found that spectra should be run at a resolution which makes the period of the noise at most half that of the bands being studied. Although this condition increases the time required to measure a spectrum with a given noise level, it makes spectral interpretation far more certain. Naturally this method cannot be used for many gaseous samples and other spectra with very narrow spectral features. In this case, the only way to ascertain whether a certain weak feature is due to signal or noise is to remeasure the spectrum using at least four times the measurement time, thereby reducing the noise level by a factor of two.

III.   EFFECT OF APODIZATION ON NOISE LEVEL

For all of the results derived above it has been assumed that the apodization function is not changed, since the magnitude of the noise level in a spectrum is dependent on the type of apodization. The effect of changing the apodization from boxcar truncation to triangular apodization nicely illustrates how the noise level is affected. For an interferogram measured with a certain retardation, the resolution (for a 100% dip between two sharp lines of equal intensity) when boxcar truncation is used in the computation of the spectrum is twice as high as if a triangular apodization function were used. However, the noise level in a spectrum computed using boxcar truncation is $\sqrt{2}$ times greater than that of the spectrum computed from the same interferogram but using triangular apodization. Thus, to measure spectra with the same noise levels from an interferogram of a given retardation, the measurement time must be doubled for the case of boxcar truncation, that is, by using twice the number of scans for a rapid-scanning interferometer or half the scan speed for a slow-scanning interferometer.

Let us compare the noise level and measurement time for spectra measured at a certain resolution, one from an interferogram measured with a retardation Δ_{max} and computed using boxcar truncation (spectrum A) and the other measured using a retardation $2\Delta_{max}$ and computed using triangular apodization (spectrum B). If the spectra were measured with the same measurement time and computed with the *same* apodization function under the constant throughput criterion, the noise level of the spectrum measured with retardation $2\Delta_{max}$ would be $\sqrt{2}$ greater than that of the spectrum measured with the lower retardation. However when the interferogram measured with the greater retardation is triangularly apodized, the noise level is reduced by a factor of $\sqrt{2}$, so that *for a given number of scans* spectra A and B will have the same noise level. Since the measurement time for B is twice that for A, and if it can be verified that the effect of the side-lobes generated with a boxcar truncation function do not have any important consequences, it is more economical to measure a spectrum with a shorter retardation and boxcar truncation.

This effect is demonstrated in Fig. 8.3 [4], where spectra measured with

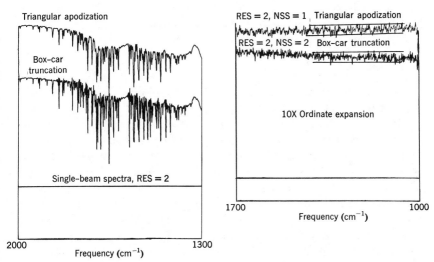

Fig. 8.3. Effect of apodization: (a) single-beam spectra computed from interferograms measured with the same retardation (RES = 2) calculated with triangular apodization (upper trace) and boxcar truncation; (b) ratio of consecutive single-beam spectra measured with the same retardation; the upper trace was measured with NSS = 1 and computed with triangular apodization, while the lower trace was measured with NSS = 2 and computed with boxcar truncation; again the noise levels are seen to be identical. (Reproduced from [4] by permission of the American Chemical Society; copyright © 1972.)

a given retardation are compared when a boxcar truncation and a triangular apodization function are used in the computations. It is seen that twice as many scans are required if boxcar truncation is used to achieve the same S/N as the spectrum computed using triangular apodization.

When the variable throughput criterion is being used, the benefits of using boxcar truncation become even more apparent, since the energy at the detector is halved for the longer retardation interferogram. Thus the measurement time must be further increased by a factor of four to achieve the same noise level as the spectrum measured with half the retardation (and twice the throughput) and boxcar truncation. This gives an overall increase in measurement time of a factor of eight. This seems to be a very large price to pay for the elimination of side lobes from a spectrum.

IV. OTHER SOURCES OF NOISE IN FTS

It should be stressed that the trading rules discussed in the previous sections are only valid if the principal source of noise in the spectrum is detector noise. There are, of course, several other sources of noise in FTS, several of which have been described in earlier sections of this book. The following paragraphs summarize the most common of these noise sources.

A. DIGITIZATION NOISE

Digitization accuracy is the principal source of noise when the signal-to-noise ratio of the interferogram exceeds the dynamic range of the analog-to-digital converter or the signal-averager. It was stated in Chapter 2 that for signal averaging to be truly valid at least the two least significant bits in the ADC should be sampling noise. If no noise whatsoever is sampled by the ADC, no improvement in S/N would be noticed for multiple scans than for a single-scan when a rapid-scanning interferometer is being used. If a slow-scanning interferometer is being used, where signal-averaging techniques are not employed, *at least* the least significant bit of the ADC should be sampling noise. It is generally true that to obtain the best results for either type of interferometer the amplifier gain should be increased to the point that the maximum amplitude of the interferogram, when digitized, is between 80 and 90% of the voltage required to fill all the bits of the ADC.

It is equally important that the word-length of the signal-averager is not exceeded for rapid-scanning interferometers. Consider the case of an interferogram whose S/N is 2^{13} being digitized with a 15-bit ADC and signal-averaged in a computer whose word-length is 16 bits. Initially the criteria for the ADC and computer word-length are obeyed. However, after 2^6 (64) scans have been signal-averaged, the S/N of the interferogram has been increased to 2^{16}, and further signal averaging will not improve this S/N unless the averaging process is carried out in double-precision (so that the effective word-length of the computer is 32 bits). This effect can cause weak absorption bands to be difficult to measure by FTS whereas weak emission bands are usually easily detected.

B.   POOR SAMPLING

If the interferogram is not sampled at precisely equal intervals of retardation, the signal that is measured is different from the signal that should have been sampled according to information theory. Therefore, an increase in the noise level in the spectrum is seen. This source of noise is illustrated in Fig. 8.4,

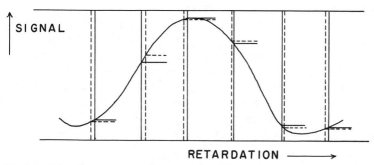

Fig. 8.4.   The effect of unequally spaced sampling points: the error (effective noise level) is dependent both on the difference between the true and the actual sampling positions and the gradient of the interferogram at that point.

where slight errors in sampling position lead to substantial errors in the measured intensity of the wave.

An uneven drive velocity of the moving mirror of the interferometer can also lead to an increase in the noise level of a spectrum. With minor variations in velocity (2% or less) the noise that is generated is negligible, but for larger variations the noise can be quite large. Of particular importance in FTS is the case where a piece of grit gets into the drive unit of a slow-scanning interferometer where the mirror is being driven by a lead screw or pulled by a cable. The mirror velocity will be reduced as the obstacle is passed and then the mirror rapidly jumps forward so that it regains its true position. In the worst case, one or two sample points can be completely missed when this effect occurs. More frequently, the modulation frequencies of the shorter wavelengths present in the spectrum exceed the time constant of the electronic filter so that incorrect amplitudes of all the component frequencies in the spectrum are measured.

It was seen in Chapter 2, Section III how missed and bad data points can also lead to an increase in noise level. The principal causes of missed data points are the reference signal dropping below the level needed to trigger the ADC (which can occur either due to the laser or a Moiré lamp failing or the reference interferometer or Moiré grating going out of alignment) or the mirror of the main interferometer jumping rapidly across several sampling positions as described above. Bad points can be measured either due to faulty electronics or when the interferometer is jarred during a measurement, which sometimes causes the signal level at the detector to change momentarily or causes a microphonic detector to pick up the vibration.

C. VIBRATION, MICROPHONICS, AND HARMONICS

Two of the detectors most commonly used in FTS are the Golay cell and the TGS detector, both of which are quite microphonic and pick up vibrations very easily. If the vibration occurs at just one point during the scan, this event gives the equivalent of a bad sampling point. However, if the vibration occurs throughout the scan and its frequency is in the range of the modulation frequencies of the spectrum being measured, when the interferogram is transformed an error (or "glitch") will be seen in the spectrum at the frequency $\bar{\nu}$ whose modulation frequency $f_{\bar{\nu}}$ (see Eq. 1.10) corresponds to the frequency of the vibration (Fig. 8.5). This effect can often be seen if vibrations from a

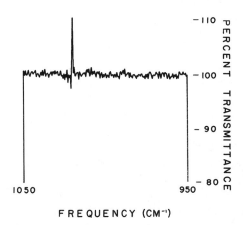

Fig. 8.5. Typical appearance of a "glitch" in the spectrum. When the scan speed of the moving mirror remains constant during the scan and the oscillation which is being picked up has a constant frequency, the glitch can be very sharp (as in this case). Variations either in the scan-speed or in the frequency of the oscillation can cause the effect to be observable over a wider frequency range.

pump attached to the interferometer or its optical bench are not well damped.

The electrical line frequency (60 Hz in the United States) often falls in the range of modulated spectral frequencies, so that if cables are not well shielded, a "glitch" may often be seen at the corresponding frequency in the spectrum. The effect is particularly prevalent if the source used in conjunction with a rapid-scanning interferometer is heated using a 60 Hz a.c. power supply. Not only will a "glitch" be seen at the fundamental frequency but also at many of the harmonic frequencies in addition. Thus well-smoothed d.c. sources have to be used with this type of interferometer.

D. FOLDING

Excess noise in the high frequency region of the spectrum can arise due to folding. It has been stressed earlier that not only should all optical frequencies above the folding frequency, $\bar{v}_{max}$, be filtered out but also the corresponding noise frequencies. Since it is very much simpler to make an optical filter with a sharp cut-off than it is to construct an electronic filter with the corresponding response, it is often found that higher frequency noise than $2V\bar{v}_{max}$ remains in the interferogram. Any information, whether it is signal or noise, which is present in the interferogram at an audio-frequency of $2V(\bar{v}_{max} + \bar{v})$ Hz is seen in the spectrum at $(\bar{v}_{max} - \bar{v})$ cm^{-1} due to folding. Thus if a filter is used which does not remove the noise above $2V\bar{v}_{max}$ Hz in the interferogram, the S/N at high frequencies in the spectrum will be degraded.

One way in which this type of noise can be avoided is to ensure that the highest optical frequencies present in the spectrum fall well below $\bar{v}_{max}$ by using a rather high sampling frequency. This type of sampling is inefficient since it requires that many more data points are collected for a given retardation than are required by sampling theory, but it does avoid the problem of folded noise at high frequencies.

E. MULTIPLICATIVE NOISE

If the intensity of a signal varies, either due to fluctuations of the source output or the detector response, noise will be seen in the spectrum over the frequency range in which the variations occur. This effect is often noticed in remote sensing studies where many factors can cause variations in the signal [6]. It is also particularly prevalent during far-infrared measurements using a slow-scanning interferometer since both the mercury lamp source and the Golay detector generally used for this work are subject to variations. For example, Golay detectors show a decrease in output signal of 1.5% for a given input signal with a temperature rise of 1°C.

No single technique will completely compensate the effects of this type of multiplicative noise. It is important that a scan speed should be used which modulates the infrared frequencies being measured outside the frequency range in which the fluctuations occur. Dual-beam techniques can be used as described in the previous chapter. The method of Thorpe et al. [7] of rapidly switching the beam between the sample and reference cells has also been shown to be effective in reducing the noise and increasing the repeatability of far-infrared measurements made with the RIIC FS–720 interferometer.

F. SUMMARY

There are obviously many sources of noise in spectra which have been measured using Fourier transform techniques. It is important to be able to recognize the conditions under which any of these noise sources could be present so that preventative measures can be taken before the spectrum becomes degraded. It is certainly true that FTS is today being used for many unusual types of measurements and will undoubtedly be used for even more in the future. Only when the principal source of noise in the spectrum is the NEP of the detector being used can the technique show results which represent its true potential.

V. INTERFEROMETERS VERSUS GRATING SPECTROMETERS

In the early days of interferometry it was not unusual to hear claims that FTS could lead to improvements of many orders of magnitude in sensitivity compared with grating spectrometers; however many of these claims were based on a poor understanding of the theory of FTS, and the real advantage of FTS over grating spectroscopy is often somewhat smaller than the unwary or overenthusiastic spectroscopist might at first think. A quantitative comparison between grating and Fourier transform spectrometers involves much more than a quick calculation of the size of Fellgett's advantage. The performance of the sources used with each type of instrument must be considered, together with other factors such as the use of choppers for modulating the signal of a slow-scanning interferometer or for double-beaming a grating spectrometer. The optical throughput, and hence the practical magnitude of Jacquinot's advantage, also has to be taken into account.

The main advantage of FTS over dispersive spectroscopy is indeed derived from Fellgett's advantage which, in quantitative terms [8], gives that the increase in sensitivity (S/N) of a Fourier transform spectrometer over a grating spectrometer with the same source, detector, optical throughput, and instrumental transmission is equal to the square root of the number of resolution elements in the spectral range being covered. Obviously Fellgett's advantage is larger for mid-infrared spectroscopy than for far-infrared spectroscopy at the same resolution and, since there are several other factors of differing relative importance between these two regions, they will be dealt with separately.

A. MID-INFRARED SPECTROSCOPY

To illustrate the important factors for mid-infrared spectroscopy, let us consider the measurement of a spectrum between 3600 and 400 cm^{-1} at 2 cm^{-1} resolution. Fellgett's advantage suggests that the sensitivity advantage of FTS, for measurements taken in equal times, is a factor of $(1600)^{\frac{1}{2}}$, that is, 40. When other factors are taken into account, however, this advantage is reduced.

For example, when the absorption spectra of samples with a high transmittance are measured by FTS, we know that a rapid-scanning interferometer must be used to avoid problems of dynamic range. At the present state-of-the-art for detection and analog-to-digital conversion, it has been determined that a fairly high scan speed (of the order of 0.4 cm of retardation per second) is required to keep the signal-to-noise ratio of interferograms of incandescent sources less than the dynamic range of a 15-bit ADC. Because of this need to maintain a high scan speed, the modulation frequencies of the short infrared wavelengths are usually greater than 1 kHz, so that the response times of the detectors commonly used in grating spectrometers (thermocouples, thermopiles, thermistor bolometers, Golay detectors, etc.) may be too slow for the high frequency components of the interferogram to be detected. The only type of wide range detector that has a sufficiently short response time to be used in conjunction with rapid-scanning mid-infrared interferometers is the pyroelectric bolometer. The sensitivity, $D*$, of the TGS pyroelectric detector is approximately equal to (and sometimes slightly better than) that of thermocouples and thermistor bolometers when low chopping speeds are considered. However, although $D*$ for a TGS detector is, to a good approximation, independent of *infrared* frequency, the sensitivity decreases as the *chopping* frequency increases. Since for a rapid-scanning interferometer each infrared frequency, $\bar{v}$, is modulated at frequency, $f_{\bar{v}}$, given by equation 1.10 as:

$$f_{\bar{v}} = 2V\bar{v} \text{ Hz;}$$

a comparison of grating and rapid-scanning Fourier transform spectrometers on the basis of detector noise shows the performance of the interferometer to be at least a factor of two down at the high frequency end of the spectrum relative to the low frequency performance.

Monochromatic radiation of intensity $I(\bar{v})$ is modulated by an ideal rapid-scanning interferometer between $-0.5I(\bar{v})$ and $+0.5I(\bar{v})$, as shown by Eq. 1.4; if an ideal grating monochromator was used to measure the same signal, a chopper is always used to modulate the beam, so that the signal would be modulated between 0 and $+I(\bar{v})$. Therefore the amplitude of the modulation using either technique is the same. On the other hand, when transmittance spectra are to be measured on an optical-null double-beam grating spectrophotometer, the measurement time for samples with a high transmittance is not increased. When an interferometer is used for the same measurement, a second (reference) interferogram must be measured separately from the sample interferogram, thereby effectively doubling the measurement time. It may be noted that this effect is partially or totally offset when optical-null grating spectrometers are used to measure the transmittance spectra of many real samples, since, for samples which transmit less than 100%, the reference beam of the spectrometer is attenuated until the signal through this beam is equal to that of the sample beam. For an interferometer, where the interferograms for the sample and reference beams are measured separately, the reference beam does not have to be attenuated in this way; thus the reference interferogram can usually be measured quickly and at a far higher signal-to-noise ratio than the sample interferogram as described in Chapter 6, Section IV. In the case of samples of low transmittance, this effect completely offsets any advantage gained by the optical-null grating spectrometer.

The efficiency of gratings and beamsplitters over extended frequency ranges also affects the relative performance of interferometers and grating spectrometers. At the center of the frequency range for which they are designed to cover, gratings and beamsplitters both have high efficiencies. The efficiency of each falls off on either side of the central frequency, and to compensate for this effect, several gratings are usually used in a grating spectrometer with an automatic grating interchange. Typical mid-infrared grating spectrometers covering the region between 4000 and 400 cm^{-1} employ three gratings which are changed during the scan with little or no loss in frequency accuracy. Like gratings, beamsplitters only cover limited ranges at high efficiency, but changing beamsplitters for mid-infrared spectroscopy is usually a slow operation at the present state-of-the-art, which helps to explain why Fourier transform spectrometers use only *one* beamsplitter to cover the same range covered using *three* gratings on a dispersive spectrometer. As noted in Chapter 4, Section II, the frequency range for greater than 80% efficiency for a typical mid-infrared beamsplitter centered at 2000 cm^{-1} ($\bar{\nu}_0$ in Fig. 4.13) is between 600 and 3400 cm^{-1}; at 3600 and 400 cm^{-1} the efficiency is about 40%. As the frequency range is extended further, for example to reach 4000 cm^{-1}, the efficiency at the extremes becomes progressively worse. If only a small frequency region at either end of the spectrum is of interest, then it is quite unlikely that interferometrically measured spectra would be superior to spectra measured on a grating spectrometer in the same time.

In view of the difficulty in aligning the beamsplitter in a Michelson interferometer, automatic beamsplitter interchanges have not yet been made for mid-infrared Fourier transform spectrometers, and this reduced performance at the extremes of the spectral range has to be tolerated. EOCOM Corporation [9] has recently developed an interferometer whose beamsplitter can be very easily changed manually, and the design of this instrument may represent the first step towards an automatic beamsplitter interchange for mid-infrared Fourier transform spectrometers. Since remote methods for aligning beamsplitters have also been described, we may not be very far away from this goal.

A further source of inefficiency of a rapid-scanning interferometer concerns the time taken to retrace the moving mirror during signal-averaging. Often the duty cycle can exceed the active scan time by as much as 30 to 300%.

Possibly the most commonly held misconception about interferometry is the magnitude of Jacquinot's advantage; statements have been often heard in the past [10, 11] to the effect that, since the area of the mirrors of an interferometer is so much larger than the area of the slits of a grating monochromator, the throughput of an interferometer must be much larger than that of the monochromator. The fallacy of this argument lies in the fact that the comparison should have been made between the area of the mirrors of the interferometer and the area of the *grating* of the monochromator (which are of approximately the same size), since the beam is focussed at a slit but is more collimated at the grating. In fact, solid angle of the beam at the grating and through the interferometer are approximately equal for measurements made at the same resolution on commercial spectrometers. It should also be remembered that when the resolution is changed, the throughput is altered by the same amount for both types of instrument.

The actual magnitude of Jacquinot's advantage for medium resolution (say 2 cm^{-1}) mid-infrared spectroscopy is still a subject of some controversy; however, the respective throughputs of commercial spectrometers are easily compared on the basis of the parameters of the beams in the sample compartments. For example, the optics of the Digilab FTS–14 have been designed so that the throughput is optimized for mid-infrared spectroscopy at 2 cm^{-1} resolution, and the area of the beam at its sample focus and the $f/$ number of the beam at this point are only slightly greater than the corresponding parameters of the Perkin-Elmer 621 grating spectrophotometer. For these two instruments at least, throughput considerations appear to favor neither the interferometer nor the grating spectrometer to any great extent. This fact makes it possible to use the same type of sampling accessories that were designed for grating spectrometers (beam condensers, ATR units, long-path gas cells, etc.) with little or no modification in the sample compartment of most Fourier transform spectrometers.

So far we have compared the time taken to measure a *spectrum* using a monochromator and an *interferogram* with an interferometer. In a valid comparison of the two types of spectrometer, the time taken for the computation and plotting of the spectrum measured using a Fourier transform spectrometer should have been taken into account.

The Cooley-Tukey FFT algorithm has reduced the computation time to the point that it is rarely now the most time-consuming step in FTS. Computing time depends first and foremost on the number of interferogram points to be transformed, but the type of computer being used and the efficiency of the programming, especially for such operations as the transfer of data from disc to core in a minicomputer, can also be very critical. For a minicomputer-disc combination, typical computation times for a single precision transform range from less than two seconds for a 1K transform to about $3\frac{1}{2}$ min for a 32K transform. Even faster rates and larger transforms can be achieved through the use of larger computers [12]. If all the computations are performed in the core of a minicomputer, 2K transforms can be carried out in less than one second. In the Spectrotherm [13] ST-10 spectrometer, a 2K transform is performed at the same time as the subsequent interferogram is being measured, thus enabling *spectra* to be signal-averaged and stored on a floppy disc.

Surprizingly enough, the stage when interferometrically measured spectra are plotted as hard copy can sometimes be the most time-consuming step in the overall measurement of a spectrum, especially for the measurement of high energy sources for which the data collection time is short. Fellgett [14] has stated that "multiplexing is like virginity; if it is to be present at the end, then it must be preserved throughout." It seems a shame that the advantage which the interferometer gains over a monochromator through multiplexing in the data collection step is reduced, and sometimes lost, through the use of a slow chart recorder.

B. FAR-INFRARED SPECTROSCOPY

Since the frequency range covered by a single beamsplitter for far-infrared spectroscopy is smaller than the range covered in mid-infrared measurements, Fellgett's advantage for a Michelson interferometer will be smaller than the factor of about 40 that was derived in the previous section for mid-infrared spectroscopy at 2 cm^{-1} resolution. For a far-infrared spectrum measured at 2 cm^{-1} resolution between 450 and 80 cm^{-1} Fellgett's advantage for FTS is approximately 14. If the interferometer is used in conjunction with a mechanical chopper (which is usually the case with slow-scanning far-infrared interferometers), this numerical factor would be reduced to about 10, since the measurement time is effectively halved.

Another factor working against the Michelson interferometer for far-infrared spectroscopy is the efficiency of Mylar beamsplitters. It was seen in Chapter 4 that the relative beamsplitter efficiency of Mylar is greater than 50% only over a very small portion of the full frequency range usually covered in practice with a single beamsplitter. For a 6.25 μm thick Mylar beamsplitter, whose peak efficiency is at 250 cm^{-1}, the efficiency is greater than 20% only between 110 and 390 cm^{-1} (for 45° incidence). Therefore for measurements made at 2 cm^{-1} resolution between 100 and 400 cm^{-1}, the advantage as measured through a comparison of S/N at the extremes of the spectral range is actually less than a factor of 2.5, although obviously the advantage around $\bar{v}_0$ is greater than this. The value of this factor may be increased through the use of several beamsplitters to cover the complete far-infrared spectrum. For lower resolution measurements over the range 400 to 100 cm^{-1}, or for measurements made at 2 cm^{-1} over lower frequency regions where the range is automatically smaller, the calculated advantage of a Michelson interferometer over a grating spectrometer after Fellgett's advantage has been offset by the low beamsplitter efficiency of Mylar beamsplitters can be negligible. It is for this reason that lamellar grating interferometers are preferable to Michelson interferometers for very far-infrared Fourier transform spectroscopy.

One important advantage of interferometry over grating spectroscopy that is particularly applicable to far-infrared spectroscopy is the ease with which stray radiation can be filtered out. This factor is particularly important for very far-infrared measurements for which second order radiation from a grating may be only a few wavenumbers higher than radiation diffracted in the first order, so that second and higher order radiation is extremely difficult to filter out without severely attenuating the first order radiation. With an interferometer however, high frequency radiation need not be filtered out provided that the sampling interval is sufficiently short that high frequency light is not folded back into the low frequency spectrum; thus all that is needed is electronic filtering to attenuate the high audio frequencies in the interferogram. It should be noted that the high frequency information outside the frequency range of interest should not be so intense that the low frequency components of the interferogram are not adequately digitized. For this reason, black polyethylene filters are usually placed before the detector of far-infrared interferometers to ensure that most of the intense mid-infrared radiation emitted from the source is removed, and further attenuation is achieved through the use of electronic filtering for both rapid- and slow-scanning interferometers.

Black polyethylene has a fairly high transmittance in the far-infrared, whereas the combined transmittance of the several filters required in a grating spectrometer is generally much lower. Among the filters that are used with a grating monochromator, the most commonly used are transmission filters, such as the Yoshinaga [15] type or alkali halide plates for very far-infrared spectroscopy, and reflection filters, such as scatter plates and *restrahlen* plates; when used in combination, these filters rarely have a transmittance above 40%. The stray light in most far-infrared grating spectrometers is still usually rather high (especially when very long wavelength measurements are being made), whereas the effect of stray light in far-infrared Fourier transform spectrometers is usually negligible.

In summary, the main advantage of interferometers for far-infrared spectroscopy from 400 to 100 cm^{-1} is derived from Fellgett's advantage, whereas for lower frequency measurements the advantage is derived principally from the energy limiting nature of the filters required in a monochromator; of course, if very far-infrared spectra are measured at high resolution, Fellgett's advantage again becomes an important factor. For both grating and Fourier transform far-infrared spectrometers, the throughput is more often than not limited by the $f/$ number of the detector fore-optics, so that the throughput of both types of instruments is similar and is not an important factor in a comparison of the performance of the two types of instrument.

REFERENCES

1. J. A. Jamieson, R. H. McFee, G. N. Plass, R. H. Grube, and R. G. Richards, *Infrared Physics and Engineering*, McGraw-Hill, New York, 1963.
2. Digilab Inc., 237 Putnam Avenue, Cambridge, Mass., 02139.
3. W. J. Potts and A. L. Smith, *Appl. Optics*, **6**, 257 (1967).
4. P. R. Griffiths, *Anal. Chem.*, **44**, 1909 (1972).
5. H. M. Pickett and H. L. Strauss, *Anal. Chem.*, **44**, 265 (1972).
6. W. J. Burroughs and J. Chamberlain, *Infrared Physics*, **11**, 1 (1971).
7. L. W. Thorpe, R. C. Milward, G. C. Hayward, and J. D. Yewen, *Optical Instruments and Techniques, 1969*, Oriel Press, Newcastle-upon-Tyne, 1969.
8. P. B. Fellgett, *J. Phys. Radium*, **19**, 187, 237 (1958).
9. EOCOM Corporation, 19722 Jamboree Road, Irvine, Calif., 92664.
10. L. C. Block and A. S. Zachor, *Appl. Optics*, **3**, 209 (1964).
11. M. J. D. Low, *Anal. Chem.*, **41(6)**, 97A (1969).
12. P. Connes, Aspen Int. Conf. on Fourier Spectrosc., 1970, G. A. Vanasse, A. T. Stair, and D. J. Baker, Eds., AFCRL–71–0019, p. 121.
13. Spectrotherm Corporation, 3040 Olcott Street, Santa Clara, Calif., 95051.
14. P. B. Fellgett, Aspen Int. Conf. on Fourier Spectrosc., 1970, G. A. Vanasse, A. T. Stair, and D. J. Baker, Eds., AFCRL–71–0019, p. 141.
15. Y. Yamada, A. Mitsuishi and H. Yoshinaga, *J. Opt. Soc. Am.*, **52**, 17 (1962).

Applications of Fourier Transform Spectroscopy

FAR-INFRARED SPECTROSCOPY

I. INTRODUCTION

Many of the earliest instrumental developments in interferometry were made in the field of far-infrared spectroscopy. The main reason that this spectral region was the first to be studied by many groups using Fourier transform techniques is the simplicity of the instrumentation required to achieve an adequate resolution for chemical spectroscopy. In the region below 400 cm^{-1}, much lower tolerances for the mirror drive of a Michelson interferometer are permitted than for the measurement of mid-infrared spectra at the same resolution, and a large optical throughput is allowed even if high resolution studies are being made. There are other additional instrumental advantages. The rather low energy of the mercury lamp sources generally used for far-infrared spectroscopy and the small spectral ranges being covered mean that the dynamic range of the interferogram is small enough to digitize easily even with slow-scanning interferometers. Also, the sampling interval required for the digitization of far-infrared interferograms is sufficiently long that less than a thousand data points are usually required for the measurement of medium resolution spectra.

It was these reasons, together with the relatively poor performance of grating spectrometers for far-infrared spectroscopy, that led to the commercial development of several far-infrared interferometers, such as those made by RIIC, Grubb-Parsons, Coderg, and Polytec. When they were initially introduced, most of these instruments did not provide any simple means of data reduction so that many chemical spectroscopists became wary of using them, and in the mid-1960s several debates [1, 2] were held on the relative merits of grating spectrometers and interferometers for far-infrared spectroscopy. To a certain extent, the debate is not yet over for several reasons, the most important being the small theoretical advantage of interferometry over grating spectroscopy for this region as discussed in the previous chapter.

Some of the factors that reduce the advantage of Fourier transform spectroscopy for far-infrared measurements are being overcome. The first drawback was the need for fast data reduction, and most manufacturers now supply a data system as part of their total package. The fall-off in the

211

efficiency of a Mylar beamsplitter on either side of its optimum frequency means that beamsplitters frequently have to be changed as different regions are examined. The introduction of the beamsplitter wheel by Polytec, which allows the beamsplitter to be changed without breaking the instrumental vacuum, permits spectra to be measured over an extended frequency range at close to the optimum beamsplitter efficiency. The loss of half the energy due to the chopper can be avoided either by the use of phase modulation techniques or by the use of a rapid-scanning interferometer. It can therefore be expected that the use of interferometry for far-infrared spectroscopy will increase rather than decrease in the next few years.

The applications of FTS for far-infrared spectroscopy can neither be defined nor summarized in a clear-cut fashion, since most chemical topics have been studied using both interferometers and grating spectrometers by different groups. The following sections are intended to summarize the important aspects of chemical far-infrared spectroscopy and to describe selected studies where Fourier transform techniques have been used to advantage. The books of Finch et al. [3] and Möller and Rothschild [4] are strongly recommended for a more complete coverage of applications of far-infrared spectroscopy.

II. INTERNAL STRETCHING AND BENDING VIBRATIONS

The absorption frequency of any vibrational mode will appear in the far-infrared region of the spectrum either if the mass of at least one of the atoms involved in the vibration is high or if the force constant for the vibration is low.

The former criterion is the one that governs the absorption spectrum of many inorganic molecules where the presence of heavy atoms generally causes bands to be seen in their far-infarared spectrum. For instance, the absorption bands of most metal oxides, sulfides, chlorides, bromides, and iodides are found below 1000 cm^{-1} and heavy metal iodides show bands below 100 cm^{-1}. Many complexes of simple salts with organic ligands produce spectra showing structure both in the mid- and far-infrared regions. The spectrum in the "fingerprint region" (above 600 cm^{-1}) is generally characteristic of the organic ligand(s) and rarely gives useful analytical information on the metal present in the complex, whereas the far-infrared spectrum usually exhibits the stretching and bending vibrations of the metal atom with the ligand.

This effect has been nicely demonstrated in a review article by Ferraro [5] who measured the spectra of a series of zinc, cadmium, and mercury halides complexed to pyrazine. It is extremely difficult to distinguish between any of these compounds from their mid-infrared spectrum, whereas the low fre-quency spectrum (in this case between 350 and 150 cm^{-1} of each complex is easily distinguishable. Ferraro makes the point that interpretation of spectra in the far-infrared can be very difficult, particularly for solids since there are so many possible bands involving metal-ligand, internal ligand, and lattice vibrations that assignment can become a real problem. Coupling of vibra-tional modes is common in the far-infrared and for complex molecules the distinction between internal vibrations and lattice vibrations can be difficult. As an analytical technique, however, far-infrared spectroscopy will allow similar complexes to be distinguished given a library of reference spectra, and Nyquist and Kagel [6] have recently published a useful compilation of spectra of inorganic compounds down to 33 cm^{-1}.

The first far-infrared spectra of inorganic compounds measured inter-ferometrically were reported by Adams and Gebbie [7] who studied several octahedral complexes of the type K_2MX_6 and square planar complexes of the type K_2MX_4, where M = Re, Os, Ir, Pt, Ru, and Pd and X = Cl, Br, and I. They showed that limited spectra-structure correlations were possible for these compounds. A more detailed study into the effects of changing a ligand on the far-infrared spectra of a series of complexes was performed by Goldstein et al. [8] using a Beckman-RIIC interferometer for their far-infrared measurements. For a series of complexes of heterocyclic bases with copper (II) halides, these authors were able to assign four bands in the far-infrared spectrum to metal-halogen stretching or bending vibrations (Fig. 9.1). Other bands in these spectra were assigned to modes involving the bond between the metal and pyridine ligand. At the time that these measurements were being made, most far-infrared spectra measured inter-ferometrically were presented in the single-beam mode, whereas nearly all spectra measured on current instrumentation would be presented in a ratio-recorded, linear transmittance, format.

Another difficulty in interpreting the far-infrared spectra of complexes is nicely illustrated by this work. The order of frequencies for the asymmetric Cu–X stretching vibration as the base is changed should reflect the changing basic strength toward the cupric halides. However, a different order is found if the asymmetric Cu–N stretching frequency is considered, an effect that had to be assigned to the tetragonal distortion caused by Cu(II) [9].

Goldstein and Unsworth have used far-infrared spectroscopy to charac-terize octahedrally coordinated halogen-bridged complexes of the first transi-tion series [10], and more recently described the far-infrared spectra of a series of dihalogeno-tetra(pyridine) complexes of Mn(II), Co(II) and Ni(II) and related six-coordinate compounds containing terminal metal-halogen bonds [11]. In all these papers, the authors used an interferometer to measure

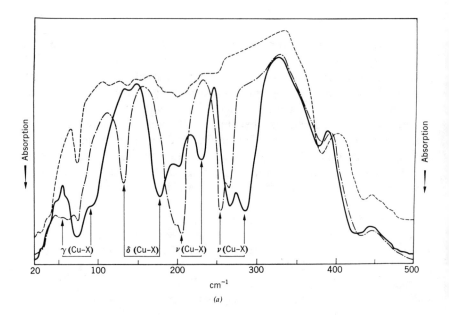

γ (Cu–X) δ (Cu–X) ν(Cu–X) ν(Cu–X)

20   100   200   300   400   500

cm^{-1}

(a)

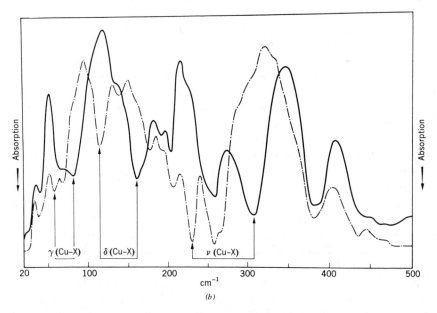

γ (Cu–X) δ (Cu–X) ν (Cu–X)

20   100   200   300   400   500

cm^{-1}

(b)

Fig. 9.1. Single-beam far-infrared spectra of some copper complexes measured with one of the first interferometers developed by Gebbie's group at the National Physical Laboratory, Teddington, England. (a) shows the spectra and assignments for $CuX_2 \cdot 2$ (pyridine) measured as a Nujol mull, where the solid line is for X = Cl and the broken line is for X = Br; the upper trace is the instrumental background; (b) shows the corresponding spectra and assignments for $CuX_2 \cdot 2$ (2-methylpyridine) (Reproduced from [8] by permission of the author and Pergamon Press; copyright © 1965.)

their far-infrared spectra and they stressed the need to extend the frequency range of the vibrational spectra of complexes below the 200 cm^{-1} limit frequently imposed by using mid-infrared grating spectrometers. They stated that previous tentative assignments of metal-halogen stretching modes involved frequencies close to, or even below, the experimental low frequency limits. Unfortunately, these assignments have been widely quoted as reference data, even though the work of Goldstein's group has proved several earlier papers to be erroneous. For example, in view of the long M–X distances for $trans$-$MX_2(py)_4$ complexes, the metal-halogen stretching frequencies are much lower than had previously been assigned with an instrumental limit of 200 cm^{-1}. A further difficulty in assigning these modes in compounds containing $terminal$ metal-halogen bonds is that they are frequently of low intensity, unlike the $v(MX)$ modes of $halogen$-$bridged$ species.

In the last few years, many inorganic chemists have begun to use far-infrared spectroscopy to characterize their compounds and to draw structural information from the spectra. Even now it is still true to say that while the metal-ligand stretching modes are usually well studied, the bending modes are rarely used for the structural interpretation of complexes. With the advent of faster, simpler far-infrared Fourier transform spectrometers, it does not seem unreasonable to expect this state of affairs to change in the future.

When a molecule is entirely composed of light atoms, the only types of vibrations that generally absorb in the far-infrared are those skeletal bending modes where more than two atoms (other than hydrogen) are involved in the vibration. For this type of vibration, even though the force constant is generally quite small, few really low frequency bands are seen in the spectra of molecules with less than eight atoms. However, one small molecule that does exhibit a fundamental absorption band below 100 cm^{-1} is carbon suboxide, $O=C=C=C=O$. The lowest frequency bending mode for this molecule was calculated to absorb near 70 cm^{-1} [12]. Smith and Leroi [13] examined the condensed phases of C_3O_2 using a Michelson interferometer and found a band at 72 cm^{-1} in the liquid sample, although no band near 70 cm^{-1} was seen in the spectrum of the solid. The intensity of combination bands involving this mode which are seen in the mid-infrared spectrum of C_3O_2 in the gas and liquid phases is also reduced after the transition to the solid phase. Smith and Leroi suggest that the amplitude of the displacement of the central carbon atom in this mode is restricted by the lattice. The same authors studied the far-infrared spectrum of carbon subsulfide [14] and surprizingly found the lowest frequency band at 94 cm^{-1} (higher than that of C_3O_2) and showed that this band can be seen in the spectra of all phases.

An important series of compounds which has several fundamental modes absorbing in the far-infrared are the substituted benzenes. For instance, in the spectra of para-disubstituted benzenes, at least five bands are generally seen below 450 cm^{-1}. These bands result from out-of-plane and in-plane skeletal vibrations of the type shown in Fig. 9.2. The force constants involved for this type of vibration are usually rather similar, and Griffiths and Thompson [15] have shown that the absorption frequencies of a series of para-dihalogeno-benzenes, X–C$_6$H$_4$–Y, can be easily correlated with the masses of the substituent atoms X and Y (Fig. 9.3). The same authors also showed that the integrated area, A, of the lowest frequency band can also be correlated with

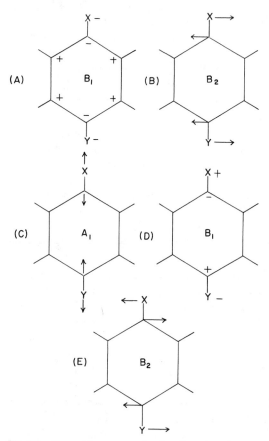

Fig. 9.2. Lowest frequency vibrational modes of para-disubstituted benzenes; all the motions involve skeletal vibrations with the heavy substituents, X and Y involved. The symmetry point group of each mode is shown for comparison with Fig. 9.3. The lowest frequency absorption band for these molecules is due to the out-of-plane bending mode (a).

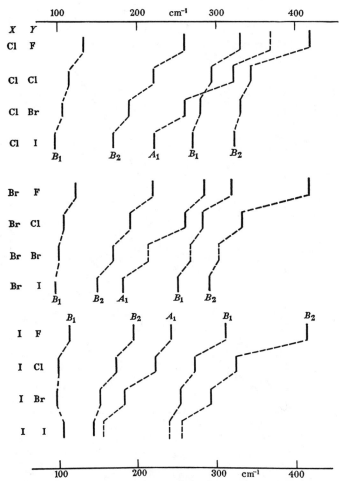

Fig. 9.3. Variation of the low-frequency bending modes of para-disubstituted benzenes with the substituents, X and Y; the heavier the substituent atoms, the further into the far-infrared the molecule absorbs. (Reproduced from [15] by permission of the Royal Society; copyright © 1967.)

the mass of the substituent atoms which suggests that the amplitude of this vibration is determined by the mass of the two substituents X and Y (Fig. 9.4). This was the first study where an interferometer was used to measure the integrated area of far-infrared absorption bands accurately; it was found that good Beer's law plots were obtained only when double-sided transforms were computed since no good phase correction routines were available at

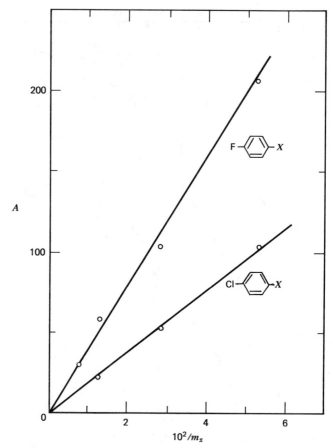

Fig. 9.4. Variation of the intensity of the lowest frequency bending mode of the para-disubstituted benzenes with the mass of the substituent, X. These integrated absorbances proved useful in assigning the absorption bands, and this study demonstrated that accurate intensity values could be measured by Fourier transform spectroscopy. (Reproduced from [15] by permission of the Royal Society; copyright ⓒ 1967.)

the time of this work. If single-sided interferograms were transformed, the zero retardation point or phase displacement had to be very accurately determined in order to obtain accurate intensity data.

For low frequency skeletal vibrational modes where many atoms are involved in the molecular motion, the absorption frequency is very sensitive to the mass of the substituent atoms. It is therefore unlikely that for such vibrations characteristic group frequencies can be defined. For some other types of low frequency vibrations, especially torsional modes, it is possible that certain far-infrared bands can be used for this purpose. Craven and Bentley [16–19] have examined the far-infrared spectra of a series of aliphatic compounds with unsymmetrical rotors, such as —OH [16], —NH₂ [17], and —OR [18]. Provided that the samples are measured as dilute solutions in cyclohexane so that hydrogen bonding effects are eliminated, the torsional bands are seen strongly at a characteristic frequency in the far-infrared. When long-chain alkyl groups are present in the molecule other skeletal frequencies are seen in the spectrum so that the assignment of a given band in the far-infrared spectrum becomes more difficult. This assignment technique cannot be applied for rotors on aromatic rings where the electron density of the bond joining the rotor to the aromatic ring can be drastically altered by other substituent groups on the ring, as described in the next section.

III. TORSIONAL FREQUENCIES

One of the first problems in physical chemistry to be attacked using far-infrared spectroscopy was the determination of the potential barrier to internal rotation about a single-bond for rotors such as —OH and —CH₃. Fateley and Miller measured the gas-phase spectra of a series of compounds with one [20], two, or three [21] methyl rotors in such compounds as $(CH_3)_2CO$, $(CH_3)_2O$ and $(CH_3)_3N$ and were able to assign torsional frequencies in the gas-phase spectrum. From the frequencies of these bands the potential barriers to internal rotation were able to be calculated and were found to agree with the value determined by microwave spectroscopy within the experimental error of both techniques [20–22]. This study was carried out with a grating spectrometer at a nominal resolution of 2.5 cm⁻¹. Tuazon and Fateley [23] remeasured the spectrum of dimethyl ether (together with those of CH_3OCD_3 and $(CD_3)_2O$) interferometrically at slightly higher resolution. The improved definition of the band envelopes measured using the Fourier transform spectrometer allowed the band centers to be located with more precision so that a value of the potential barrier could be obtained in closer agreement with the microwave value. The spectra of dimethyl ether with varying degrees of deuterium substitution as published by Tuazon and Fateley in 1972 are shown in Fig. 9.5.

The potential barrier to internal rotation of groups attached to aromatic systems is of great interest to both physical chemists and organic chemists. The variation of this barrier as other substituent groups on the ring are changed should give important information on the change of electron density at the single bond about which the vibration is occurring.

The torsional band of substituted benzaldehydes in the liquid phase was

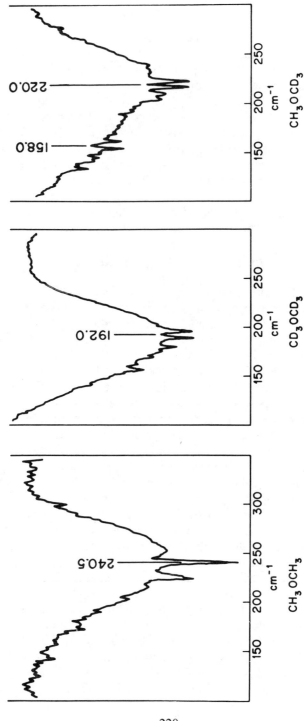

Fig. 9.5. Far-infrared absorption spectra of three deuterated dimethyl ethers measured in the gas phase (10-cm path length) using a Digilab FTS–14 spectrometer. (Reproduced from [23] by permission of the author and the American Institute of Physics; copyright © 1971.)

220

first studied by Silver and Wood [24]. Fateley et al. [25] later showed that a large shift occurred for the torsional frequency on changing from the liquid to the gaseous phase, and stressed that only the gaseous spectra should give meaningful results. Wood's work showed that the lowest frequency band in the far-infrared spectrum was strongly dependent on the other substituents present on the ring. However, the calculated potential barrier was difficult to correlate with the expected inductive and mesomeric effects of the substituent groups, especially with respect to the para-substituted benzaldehydes. This effect is almost certainly caused by coupling with the lowest frequency out-of-plane bending mode of the para-substituted benzene ring.

One type of aromatic molecule where this kind of coupling would be expected to be much smaller is the phenols, and these molecules have been subjected to an extensive survey by Fateley and Carlson [26, 27] using a rapid-scanning interferometer. Rather than measuring gas-phase spectra with the associated energy losses of the long-path gas-cells necessary to measure sufficiently strong absorption bands of these relatively nonvolatile compounds, Fateley and Carlson were able to use dilute solutions in cyclohexane. In an earlier study Carlson et al. [28] showed that torsional frequencies measured in dilute cyclohexane solution absorbed at almost exactly the same frequency as the corresponding gas-phase spectra. In the same paper, these authors described a simple way of deuterating phenols for far-infrared studies, so that both the —OH and the —OD forms of the phenol could be easily studied. The far-infrared spectrum of o-cresol in cyclohexane solution is shown in Fig. 9.6.

Fateley and Carlson found that the potential barrier for internal rotation of the —OH group in the phenols could be very easily correlated with the Hammett σ_p and σ_m parameters which have been used to predict inductive and mesomeric effects in substituted benzenes. These authors believe that the potential barriers calculated from their far-infrared data might actually lead to a more accurate value of the Hammett parameters than the ones derived from the ionization constants of substituted benzoic acids. The σ_p and σ_m values calculated by Fateley and Carlson agree well with the a priori values calculated by Radom et al. [29].

IV. PUCKERING MODES IN SMALL RINGS

Another type of vibrational mode which can give direct information on a potential function is the ring puckering mode of small ring compounds. To obtain useful information on the potential function by far-infrared spectroscopy, fairly high resolution spectra of the molecule in the gas phase have

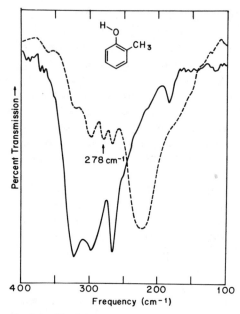

Fig. 9.6. Far-infrared spectra of ortho-cresol (solid line) and ortho-cresol-OD in cyclohexane solution. Spectra were measured in a 5-mm cell with polypropylene windows, and the sample spectra were ratioed against a compensating cell with pure solvent. (Reproduced from [28] by permission of the author and Pergamon Press; copyright ©1972.)

to be measured in order that a series of Q branches may be observed corresponding to the $\Delta n = 1$ transitions between the pseudorotational energy levels.

Several research groups are currently actively interested in studying this type of molecule and four-, five-, and six-membered rings have all been investigated. The ring puckering vibration of the parent hydrocarbon in each series (cyclobutane, cyclopentane, and cyclohexane) is infrared inactive. Thus, in order that the ring puckering transitions become active in the far-infrared spectrum either one or more of the hydrogen atoms on the ring must be substituted by an electronegative atom such as fluorine or oxygen (e.g., cyclobutanone) or one of the methylene groups in the ring is replaced by another atom or group such as O, S, or NH (e.g., tetrahydrofuran).

The first molecule for which pseudorotation was verified through its far-infrared spectrum was tetrahydrofuran [30]. This spectrum was remeasured and confirmed using an interferometer by Greenhouse and Strauss [31], who at the same time measured the spectrum of another five-membered ring, 1, 3 dioxolane. The potential barrier of each of these molecules was found to be approximately equal, having a values close to 50 cm^{-1}.

Durig and Lord [32], using a grating spectrometer, and Borgers and Strauss [33], using an interferometer, published the far-infrared spectra of several four-membered ring compounds in 1966. Both groups measured the spectra of cyclobutanone and trimethylene sulfide, and it is interesting to compare the quality of the spectra produced by these two instruments at approximately the same resolution (Fig. 9.7). The Beckman IR-11 grating spectrometer and the RIIC FS-520 interferometer were representative of the state-of-the-art for commercial far-infrared spectrometers at the time of these measurements; both instruments used a mercury lamp source and Golay pneumatic detector, so that these spectra are indicative of the relative performance of the instruments under reasonably high resolution conditions. Borgers and Strauss reported several features not able to observed with the grating spectrometer, and subsequent work has verified that these bands are indeed real. Many other small ring compounds have since been measured by far-infrared spectroscopy using both grating and Fourier transform spectrometers which have yielded useful information on the conformation and potential function for this type of molecule.

V. PURE ROTATIONAL SPECTRA

The previous two sections have covered torsional modes (internal rotation) and ring puckering modes (pseudorotation); it therefore seems fitting to cover pure rotation spectra in this section. The pure rotational spectra of heavy molecules are measured in the microwave region of the spectrum. As the molecular weight of a molecule decreases the frequency of the rotational transitions increases, and the pure rotational spectra of many small molecules can be measured using far-infrared spectrometers.

One particular rotation spectrum, that of water vapor, is only too well known to all far-infrared spectroscopists! Owing to the strength of the pure rotation spectrum of water vapor, all far-infrared spectrometers must be evacuated (or at least very well purged) so that the source energy is not too severely attenuated by these absorption lines before reaching the detector. The water vapor spectrum can also be used as a test of resolution, and many spectroscopists have demonstrated the performance of their spectrometer by resolving certain lines in this spectrum. For example, Sanderson and Scott [34] demonstrated that the resolution of the high-resolution Michelson interferometer which they constructed was considerably better than 0.1 cm^{-1} by completely resolving two bands around 59.9 cm^{-1} which are 0.093 cm^{-1}

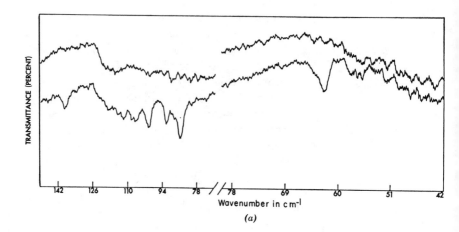

(a)

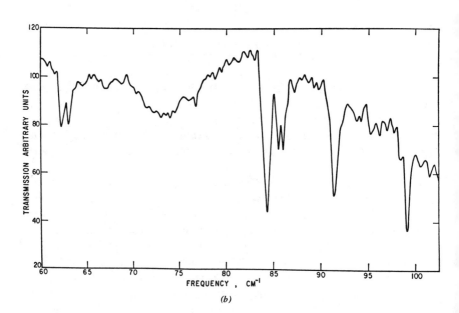

(b)

224

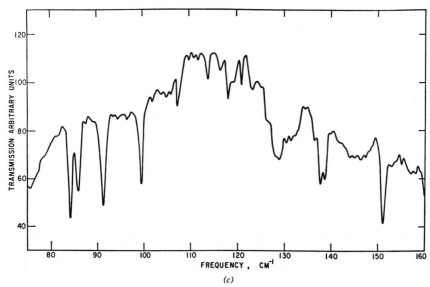

(c)

Fig. 9.7. Far-infrared spectra of trimethylene sulfide in the gas phase: (a) spectra measured using a Beckman IR–11 grating spectrophotometer with an 8.2-m cell with polypropylene windows. The upper curve is the spectrum of the empty cell with two pieces of polypropylene in the reference beam. The lower curve is of the cell containing trimethylene sulfide at 8 torr. (Reproduced from [32] by permission of the author and the American Institute of Physics; copyright ⓒ 1966.); (b) spectra measured using a RIIC FS–520 interferometer with a 1-m cell with high-density polyethylene windows. The pressure is stated to be between 20 and 35 torr. The upper spectrum is measured using a 25 μm thick Mylar beamsplitter and the lower spectrum with a 12.5 μm film. (Reproduced from [33] by permission of the author and the American Institute of Physics; copyright ⓒ 1966.)

apart (Fig. 9.8*a*). The same two regions of the spectrum were used by Hall, Vrabec and Dowling [35] to demonstrate the performance of their lamellar grating interferometer, and their spectra are shown in Fig. 9.8*b*.

For linear molecules, the rotational constant B_o and the centrifugal distortion constant D_o are easily derived from the far-infrared spectrum. Among the first diatomic molecules whose far-infrared spectrum was measured interferometrically was carbon monoxide [36], and this spectrum is now known so accurately that it is commonly used for the frequency calibration of all types of far-infrared spectrometers. Hall and Dowling [37] have also measured the pure rotation spectrum of nitric oxide which is of interest since NO is a stable diatomic molecule having an odd number of electrons. The ground state is a $^2\Pi$ state which is split into $^2\Pi_{1/2}$ and $^2\Pi_{3/2}$ components by spin-orbit interaction. The splitting is sufficiently small (122 cm^{-1}) that there is an appreciable population of both components at room temperature. The

H2O

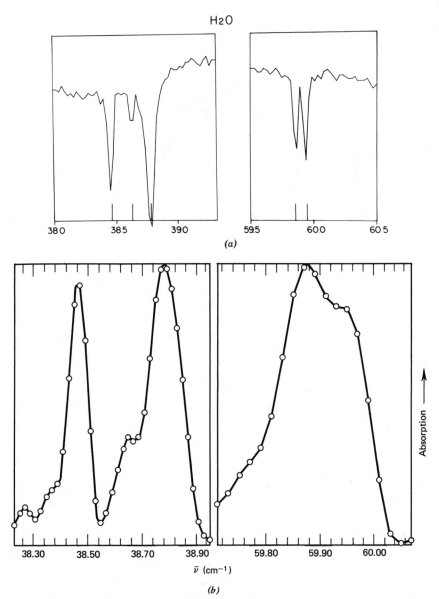

Fig. 9.8. Pure rotation spectra of water vapor measured to demonstrate the resolution of two Fourier transform spectrometers: (a) spectra from a high resolution Michelson interferometer with a maximum retardation of 17.2 cm. (Reproduced from [34] by permission of the author.); (b) spectra measured with a lamellar grating interferometer whose maximum retardation was 16.0 cm. (Reproduced from [35] by permission of the author and the Optical Society of America; copyright © 1966.) From a comparison of these two spectra it appears that the theoretical resolution of the lamellar grating interferometer was not being attained, although with the pressure of the water vapor being used for this measurement there may have been some pressure broadening of the absorption bands.

pure rotation spectrum of a normal linear diatomic molecule consists of a set of equally spaced lines so that the interferogram shows a series of "signatures" whose separation is inversely proportional to the separation of the absorption lines in the spectrum. Since the NO spectrum consists of two series of lines (corresponding to the $^2\Pi_{1/2}$ and $^2\Pi_{3/2}$ states), the signatures corresponding to each series are seen in the interferogram and became better resolved as the retardation is increased (Fig. 9.9). The spectrum found on

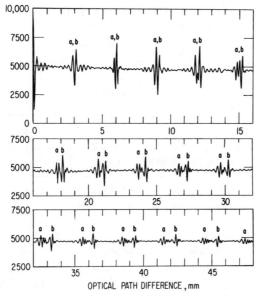

OPTICAL PATH DIFFERENCE , mm

Fig. 9.9.   The interferogram of nitric oxide showing two sets of signatures (marked a and b) from the two $^2\Pi$ states; this spectrum was measured with the same lamellar grating interferometer used for Fig. 9.8b. (Reproduced from [37] by permission of the author and the American Institute of Physics; copyright © 1966.)

transforming this interferogram is seen in Fig. 9.10. Another fact that Hall and Dowling deduced from this interferogram was that, since the adjacent signatures in each set are nearly inverted copies of each other, the quantum number J in the energy level expression is indicated to be non-integral. From their spectra, Hall and Dowling were able to measure B_o to an accuracy of $\pm 0.005\%$ for each state, D_o to an accuracy of $\pm 2\%$ in one case and even found a higher order distortion constant, H_o, accurate to $\pm 25\%$.

Another set of molecules whose far-infrared rotation spectrum is fairly

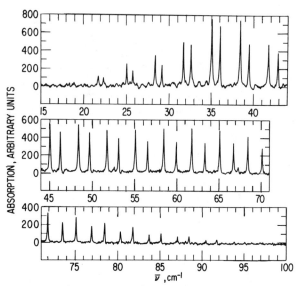

Fig. 9.10. The transform of the interferogram shown in Fig. 9.9 showing two series of pure rotation lines resulting from the $^2\Pi_{1/2}$ and $^2\Pi_{3/2}$ states of nitric oxide. (Reproduced from [37] by permission of the author and the American Institute of Physics; copyright © 1966.)

easily interpreted are the C_{3v} symmetric tops. The spectra of a series of prolate symmetric tops including the methyl halides and methyl cyanide and their totally deuterated analogs and the methyl acetylenes $CH_3C\equiv CH$ and $CH_3C\equiv CD$ were measured interferometrically by Griffiths and Thompson [38]. Since these molecules can be treated in the same fashion as linear molecules, the rotational constants B and D_J were easily derived from the spectra. For oblate rotors (whose moment of inertia about the C_3 axis is greater than the other two moments) the intensity of the K substructure in each J line increases with K so that, unlike the case of the prolate tops, the intensity of each J line is not concentrated near $K = 0$. Consequently a correction to allow for this effect had to be made in order that the rotational constants derived from the far-infrared spectrum agree with those measured from the microwave spectrum. In this way the spectrum of fluoroform (a typical oblate rotor) could be accounted for; however, the spectrum of a more complex oblate top, trimethylamine, showed further anomalies. This molecule is sufficiently heavy that the intensity of rotational lines above 40 cm^{-1} should be negligibly small. However strong lines are seen in the higher frequency region of the spectrum (Fig. 9.11), the spacing of which is

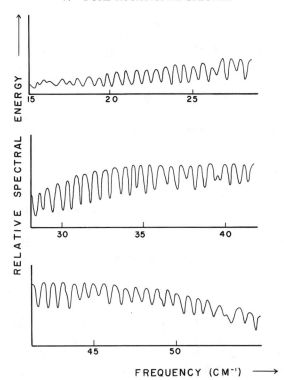

Fig. 9.11. The single-beam pure rotation spectrum of trimethylamine measured with a RIIC FS–520 interferometer. For similar molecules with this symmetry and geometry the absorption of the pure rotation lines would be very low above 40 cm^{-1}; higher frequency lines are a result of coupling with the internal rotation of the methyl groups.

not found to correspond with the pure rotation spectrum at lower frequency. Like the interferogram of NO, the interferogram of trimethylamine shows two signatures appearing quite close together. The high frequency lines probably cause the second signature, which seems to be due to coupling of the pure rotation spectrum with internal rotation of the methyl groups in $(CH_3)_3N$.

The pure rotation spectra of several asymmetric rotors have been measured interferometrically, including those of O_3, SO_2 [39], and methanol [40]. Each of these spectra look rather similar to the rotation spectra of symmetric tops because two of their principal moments of inertia are very similar and quite different from the third. Gebbie et al. [39] have analyzed the spectra of O_3 and SO_2 carefully and have calculated the rotational constants which are in good agreement with the microwave values. These spectra are important since Harries et al. [41] have shown that far-infrared rotation spectra could be used in pollution and stratospheric studies. This technique has the disadvantage that transmission through "windows" in the rotational spectrum of water vapor must be used for ground-level measurements, but on the positive side, far-infrared radiation is far less susceptible to scattering by fog or solid particulate matter in view of the longer wavelengths involved [42].

VI. COLLISION-INDUCED SPECTRA

Collision-induced absorption in the region of rotational transitions was first observed for hydrogen in the mid-infrared region of the spectrum [43]. For hydrogen at high pressure the spectrum shows the pure rotation bands (which are infrared inactive at low pressure) superimposed on a continuum. The continuum was believed to be associated with a process in which two molecules in collision increase their kinetic energy directly by absorption of a photon without any change in their internal energy [44, 45]. This theory was confirmed by using compressed mixtures of rare gases. The mid-infrared measurements generally indicated that the maximum absorption should occur in the far-infrared. Bosomworth and Gush [46, 47] constructed a Michelson interferometer for the purpose of studying these collision-induced bands in the far-infrared, and showed the existence of the band for mixtures of several different gases. For example they demonstrated that the translational spectrum of a neon-argon mixture had a maximum absorption close to 100 cm^{-1} while a helium-argon mixture had its peak around 190 cm^{-1} (Fig. 9.12).

When homonuclear diatomic molecules are studied at high pressure in the far-infrared, the rotational spectrum may absorb in the same frequency range as the translational spectrum. This is the case for oxygen and nitrogen but for hydrogen the pure rotational spectrum is seen at higher frequency than the translational spectrum. Gebbie et al. [48] first measured the spectrum of compressed nitrogen, and Bosomworth and Gush [47] remeasured the spectrum to obtain more precise intensity data. The far-infrared band in the compressed nitrogen spectrum arises predominantly from dipole moments which result from a quadrupole induction mechanism. As a result, the quadrupole moment may be deduced from the integrated intensity of the band. The values of the quadrupole moments of oxygen and nitrogen calculated from the spectra of Bosomworth and Gush agree well with values calculated by other methods.

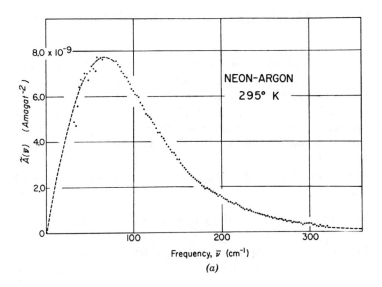

(a)

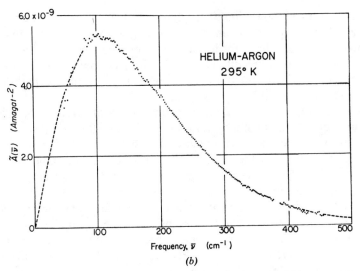

(b)

Fig. 9.12. Translational spectra of compressed mixtures of (a) neon and argon and (b) helium and argon, measured using the Michelson interferometer designed by Bosomworth and Gush [46]. The dots are experimental points and the dashed lines are extrapolations. (Reproduced from [47] by permission of the National Research Council of Canada; copyright ©️ 1965.)

231

VII. NONSPECIFIC FAR-INFRARED ABSORPTION IN LIQUIDS

Several reports have shown that the far-infrared spectra of polar liquids show a strong, broad absorption feature below 100 cm^{-1} which cannot be attributed to intramolecular vibrations or pure rotations. Microwave studies have shown that strong absorption occurs *below* the region that intramolecular absorption bands are observed, due to Debye-type relaxation, but neither mechanism can account for all the absorption above 20 cm^{-1}. Similar but much weaker far-infrared absorption has been observed in the spectra of nonpolar liquids.

Various explanations have been proposed as to the origin of these absorption bands.

1. The bands arise from the stretching modes of a dipole-dipole complex [49].
2. They are due to vibrations of a "liquid lattice" [50].
3. The bands arise from a hindered rotation of the molecule in a cage of its neighbors [51].

The existence of dipole-dipole complexes was made rather unlikely by the observation [52, 53] that the intensity of the bands of such polar molecules as dimethyl sulfoxide, acetonitrile and acetyl chloride does not decrease even when these molecules are diluted to 0.2% in a nonpolar solvent such as cyclohexane, provided that the product of concentration and path-length is held constant (Fig. 9.13).

Chantry and Gebbie [50] compared the spectra of several molecules in the liquid and crystalline phases and observed that while the absorption in the liquid phase spectrum is broader and at slightly lower frequency than the lattice vibrations in the crystal phase spectrum of the same molecule, there is a noticeable correspondence between the low frequency bands in these spectra. This observation led them to postulate that the liquid can be considered as a disordered crystal lattice, so that the same factors leading to lattice vibrations would lead to absorptions in the liquid phase spectrum. Figure 9.14 shows liquid and solid phase spectra of chlorobenzene measured by Chantry and Gebbie [50]; further results of this nature have been published by Fleming et al. [54].

These nonspecific far-infrared absorptions appear to be best explained by the hindered rotation model of Kroon and van der Elsken [51], and Davies and his co-workers [55] were able to explain the band parameters of far-infrared absorptions of polar liquids not exhibiting hydrogen bonding using this model. Davies et al. [55] measured the spectra down to 2 cm^{-1}

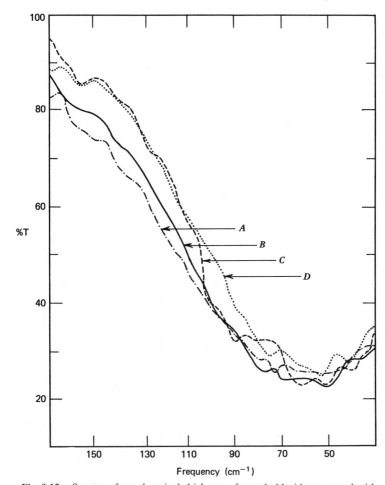

Fig. 9.13.   Spectra of equal optical thickness of acetyl chloride measured with
an RIIC FS–520 interferometer: (a) Pure liquid, 0.05 mm thick; (b) 50% solution
in cyclohexane, 0.1 mm thick; (c) 10% solution in cyclohexane, 0.5 mm thick;
(d) 5% solution in cyclohexane, 1.0 mm thick; (e) 1.6% solution in cyclohexane,
3.0 mm thick.

using an NPL "cube" interferometer with an indium antimonide detector
at 1 K for their very far-infrared studies, so that they were able to overlap
their results with spectra measured using microwave techniques. Davies
and Chamberlain [56] have stressed the need for high photometric accuracy
and repeatability for these measurements and they have published spectra

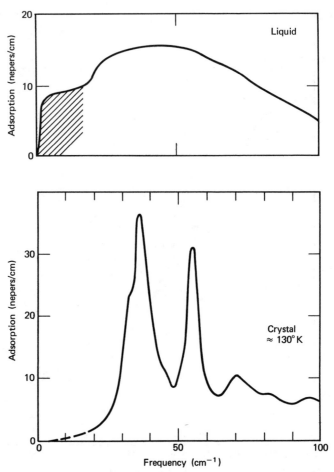

Fig. 9.14. Spectra of liquid and solid chlorobenzene measured with an NPL "cube" interferometer, showing how the spectrum of a polar liquid is rather similar to a "smeared-out" spectrum of a crystal of the same material. Below 20 cm^{-1} (shaded region of the liquid spectrum) Debye relaxation processes are dominant for the liquid. (Reproduced from [50] by permission of the author and MacMillan (Journals) Ltd,; copyright ©️ 1965.)

of *p*-difluorobenzene demonstrating these properties. Figure 9.15 shows three independent measurements of the power absorption coefficient of *p*-difluorobenzene in carbon tetrachloride, which show a strong intramolecular vibration band at about 170 cm^{-1} and the somewhat weaker lower frequency band centered close to 70 cm^{-1} which is characteristic of polar liquids.

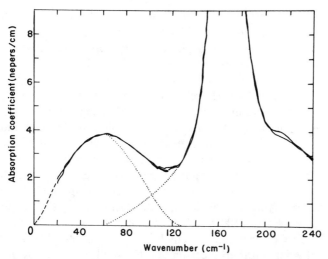

Fig. 9.15. Three independent determinations of the power absorption coefficient of a 6.16 M solution of p-difluorobenzene in carbon tetrachloride measured using the NPL–Grubb Parsons cube interferometer to demonstrate the repeatability of far-infrared spectra which can now be attained using a Michelson interferometer. (Reproduced from [56] by permission of the author and the Institute of Physics; copyright © 1972.)

VIII.   WEAK INTERMOLECULAR INTERACTIONS

In the discussion of internal stretching and bending vibrational modes, it was stated that if the force constant for the vibration is small, the frequency of the absorption band corresponding to that fundamental mode will be low. Thus for organic molecules with strong covalent bonds, most vibrational modes absorb in the mid-infrared region of the spectrum and only skeletal bending modes absorb in the far-infrared. For complexes with weak *inter-molecular* bonds such as hydrogen bonds and charge-transfer interactions, the force constants for stretching and bending of the intermolecular bond are much lower than those for covalent bonds, so that the fundamental stretching and bending modes of the complex absorb in the far-infrared.

The first interferometric study of hydrogen bonding in the far-infrared was performed by Lake and Thompson [57] on a series of straight-chain aliphatic alcohols in both the liquid and the solid state. The hydrogen bond stretching band varied between 110 cm^{-1} and 219 cm^{-1} for these molecules. Careful distinction had to be made between the hydrogen bond stretching frequency and the —OH torsion of the monomeric alcohols in solution, which also absorbed in the region of 200 cm^{-1}. The torsional band was found to shift markedly on changing the environment of the molecule. In cyclohexane solution, the frequency of the torsion corresponded closely with the frequency measured in the gas phase. On the other hand, when benzene was used as a solvent the torsional frequency shifted to about 300 cm^{-1} and in the undiluted alcohols it was found around 600 cm^{-1} showing that the torsional mode of the —OH group is very sensitive to hydrogen bonding.

A large amount of additional work has been carried out on the far-infrared spectroscopy of hydrogen bonded species in solution, although most of the published work has involved measurements made with grating spectrometers.

Thomas [58–60] has studied the far-infrared spectra of binary mixtures of gaseous molecules which might be expected to exhibit strong hydrogen bonding. Earlier work using mid-infrared spectroscopy had demonstrated that the spectra of mixtures of HF and ethers show a broad absorption band in the range of 3900 to 3100 cm^{-1} corresponding to the stretching vibration of HF shifted to lower frequencies by hydrogen bonding [59]. Thomas [60] measured the far-infrared spectrum of HF and DF with ROR' (where R and R' can be methyl or ethyl) using an NPL "cube" interferometer, and observed a band at 180 cm^{-1} which he assigned to the intermolecular stretching vibration of the hydrogen bonded complex. The width of this band was able to be correlated with the width of the corresponding combination band in the mid-infrared spectrum. The frequency shifts for the deuterated species and calculations of the change in vibrational entropy were consistent with the assignment of the observed bands to the hydrogen bond stretching vibration.

One very interesting paper by Gebbie et al. [61] has suggested that a dimer of water exists in the gas phase and evidence for this dimer can be seen in the very far-infrared spectrum of water vapor at long path length. In 1967, Viktorova and Zhevakin [62] predicted that the dimer could exist and that it would show a discrete absorption band at 7.1 cm^{-1} and a continuum at higher frequencies. This continuum could account for the excessive absorption observed in the very far-infrared atmospheric "window" regions. The continuum is thought to consist of a large number of weak features due to rotational transitions both in the vibrational ground state and excited states of the dimer. Using an NPL-Grubb Parsons modular interferometer, Gebbie et al. [61] measured the absorption spectrum of pure water vapor at various temperatures and pressures, and observed a discrete feature at 7.5 cm^{-1}. The strength of this bond is proportional to the square of the pressure, indicating that the absorption is caused by the interaction of pairs of water molecules. An Arrhenius plot shows the binding energy of each dimeric pair to be 6 ± 3 k-cals/mole. Weaker absorption lines were detected in the vicinity of 9.3 cm^{-1} and 22 cm^{-1} but they were of very low intensity. Experiments using an HCN maser at two different far-infrared frequencies verified the existence of the continuum. Solar spectra show both the 7 cm^{-1} and 9 cm^{-1} features and the anomolies in the spectral windows.

The implications of this work are quite far ranging. For instance, in meteorology the newly found atmospheric constituent must be included in radiation balance and heat exchange calculations for the atmosphere. Gebbie et al. [63] have suggested that the very far-infrared spectrum can be used to determine the concentration of stratospheric water vapor. In their paper on the water dimer, Gebbie and his co-workers [61] suggest that it should also be possible to measure dimeric water absorption as an independent parameter, and because this band has a different temperature and pressure dependence than the pure rotational bands of the monomer, data on the pressure and temperature of atmospheric water vapor should be able to be obtained from a single spectrum.

Another type of weak intermolecular interaction to have been studied by far-infrared spectroscopy is the bonding in charge-transfer complexes. Ginn and Wood [64] first demonstrated the existence of far-infrared absorption bands due to the intermolecular stretching vibration of a series of charge-transfer complexes of pyridine with I_2, IBr, and ICl. Lake and Thompson [65] then measured the far-infrared spectra of cyclohexane solutions of charge-transfer complexes of substituted pyridines with I_2 and Br_2 using an RIIC FS–520 interferometer. Bands observed in ten compounds for the iodine complex were assigned to the I–I stretching vibration (near 180 cm^{-1}) and the I-pyridine stretching vibration (between 95 and 65 cm^{-1}). Equilibrium constants for the formation of the complex were estimated by dilution studies. In a later study of the same system using a grating spectrometer, Yarwood and Person [66] found slightly different values for the equilibrium coefficient and assigned a possible reason for this difference to the stray light in their spectrometer.

IX. INTERMOLECULAR VIBRATIONS OF MOLECULAR CRYSTALS

Whenever spectra of solids are measured in the far-infrared, absorptions due to lattice vibrations will be seen. In crystals of light molecules there is usually a fairly clear distinction between the internal vibrations and the lattice vibrations, since the forces connecting atoms within the molecule are generally stronger than the forces acting between the molecules. Since the number of molecules in the unit cell determines the number of lattice vibrations a large number of vibrational modes are possible. However, the theory of crystal dynamics shows that the only modes which are infrared active are those for which the wave vector, k, associated with the vibration is equal to zero, such that all molecules are oscillating with the same frequency and phase. To a good approximation each mode of vibration for which $k = 0$

can be described as a translation or a rotation (or libration). Thus, if the symmetry of the crystal and the number of molecules per unit cell are known the number of infrared active modes can be calculated. Conversely, if the crystallographic data are not available, the number of experimentally observed bands assigned as lattice vibrations can provide information on this subject.

Since lattice vibrations of molecular crystals usually appear in the very low frequency region of the spectrum, experimental difficulties have meant that few spectra of molecular crystals were measured until the development of the interferometer. Anderson et al. [67–73] have investigated the far-infrared absorption spectra of a number of molecular crystals at liquid nitrogen temperature. This group has measured the spectra of crystalline HCl and HBr [67] and their deuterated analogs [68], the halogens Cl_2, Br_2, and I_2 [69], the hydrides H_2S and NH_3 and their fully deuterated analogs [70], and other small molecules such as CO_2, N_2O, and COS, [71] SO_2 [72], and C_2H_2 and C_2D_2 [73].

Frequency shifts on deuteration of molecules containing hydrogen have been used to distinguish between translational and librational modes, since bands that shift by a factor of $\sqrt{2}$ can be assigned to librational modes, whereas those that do not shift very much are due to translations. Anderson et al. were able to verify the crystal structure of materials whose crystallographic data were known and perform normal coordinate analyses for these compounds including nearest-neighbor interactions in the computations. Occasionally, broadening of a band would be observed in the spectrum and this effect was attributed to orientational disorder.

X. IONIC CRYSTALS

Ionic crystals can be classified into two types—those containing groups of covalently bonded atoms such as oxyanions or complex ions and those in which each ion consists of a single atom. From the point of view of molecular spectroscopy, the former group can be treated in a similar fashion to the spectra of molecular crystals while the pure ionic crystal has to be treated differently. The theory for this type of crystal has been described simply in articles by Mitra [74] and Martin [75].

Little work in this field has been performed using Fourier transform techniques and so the subject will be only briefly summarized. Many detailed calculations have been made using the absorption and reflection spectra of pure ionic crystals. Careful analysis has also been carried out on the far-infrared spectra of the crystals of salts containing simple oxyanions, such as the nitrate ion, and force constant data can be deduced from these calculations. For complex ions, most workers have been interested only in the internal vibrations and they have generally merely assigned low frequency bands to lattice vibrations without any further analysis.

XI.   LOW TEMPERATURE SPECTROSCOPY

In many applications of far-infrared spectroscopy, useful information can be obtained by reducing the temperature of the sample. In particular, temperature reduction has been suggested as a useful method for distinguishing between intramolecular vibrations and lattice modes, since generally the frequency of lattice bands increases on cooling the sample while the absorption frequency of fundamental modes is usually unaffected.

The first study where an interferometer was used in conjunction with a cryostat for chemical spectroscopy was described by Lake and Thompson [57] who cooled a series of aliphatic alcohols down to the temperature of liquid nitrogen in an attempt to assign the hydrogen bond stretching frequency, and they were able to assign this mode for most of the compounds which they studied. In the cryostat used by Lake and Thompson, the sample was held between two polyethylene windows clamped to the bottom of the cryostat. The interior of the RIIC FS–520 interferometer was evacuated before the cryostat was cooled so that the interferometer vacuum also acted as the Dewar vacuum. In this way, no further windows were needed and the energy reaching the detector was approximately the same as for ambient temperature measurements.

Griffiths and Thompson [15] measured the spectra of a series of para-disubstituted benzenes at liquid nitrogen and ambient temperature. To eliminate lattice vibrations the samples were dispersed in a polyethylene matrix after the method of Brasch and Jakobsen [76]. Discs of low density polyethylene sheet were heated until the crystallites just melted and a small quantity of the sample was sprinkled on the surface so that it dissolved in the molten polyethylene. On cooling the disc back to room temperature the sample molecules are dispersed throughout the polyethylene matrix and no intermolecular vibrations are observed. This study showed that the lowest frequency out-of-plane bending mode of the molecules being measured was quite temperature-dependent, so that care has to be used when applying the rule that *only* lattice modes are temperature dependent.

As stated earlier, the cryostat used for these measurements had a very high transmittance; however the cryostats used for some other low temperature measurements can cause much greater energy losses and the use of interferometers becomes even more advantageous. One of the most important measurements falling into this category involves the spectra of molecules dispersed at low concentration in a condensed inert gas matrix.

Matrix isolation techniques are often used for the spectroscopy of unstable or reactive species, but few far-infrared measurements have been performed because of transmission losses in the cryostat further limiting the energy in an already energy-limited spectral region. Considering the very low temperatures needed to condense the inert gas molecules forming the matrix, the spectrometer vacuum can no longer be used as the Dewar vacuum and the sample area must be separately evacuated and held between two thick transmitting windows (polyethylene is usually used for far-infrared spectroscopy). Since the gases are usually condensed on a silicon plate (causing further energy losses through reflectance) only about 20% of the normal source energy generally reaches the detector.

Barnes et al. ⌊77⌋ reported the first interferometric measurement of species trapped in a low temperature matrix. They studied the rotation of some hydrogen and deuterium halides in argon and nitrogen matrices using an RIIC FS–720 interferometer. For HCl, DCl, and HBr in argon, Barnes et al. observed absorption bands at 18.6, 10.9, and 16.2 cm^{-1}, respectively, which they were able to attribute to the $J = 0 \rightarrow 1$ rotational transition of the halides. No similar bands were observed when the same molecules were trapped in a nitrogen matrix, confirming earlier conclusions which had been drawn after a study of the mid-infrared spectra of the same systems.

In a more recent article, Guillory and Smith [78] have described measurements made using a Digilab FTS–16 far-infrared Fourier transform spectrometer for the matrix isolation spectroscopy of hydride and halide germanium compounds in an argon matrix. Most of the samples were diluted in the ratio of 400 parts of argon to one part of the sample. A particularly interesting spectrum shown in this article concerned the photolysis of GeH_2Br_2. Guillory and Smith were able to show the presence of three new bands in the spectrum after the matrix was photolyzed for 5 hr with a medium-pressure mercury lamp (Fig. 9.16). These measurements not only demonstrate the instrumental performance in terms of signal-to-noise ratio but in addition illustrate the advantage of having the data in a digital format. The first two spectra show the transmittance of the matrix before (a) and after (b) photolysis. The third spectrum (c) is the ratio of the two previous spectra and shows much more clearly the absorption bands (downward-going) of the newly formed species. Of these three bands, the strong one, at 276 cm^{-1}, is easily seen in spectrum (b) whereas the two weak ones are much more difficult to observe. These spectra demonstrate how much easier it is to observe weak bands on a flat background than on a rapidly changing baseline. Guillory and Smith have demonstrated that very good spectra of highly diluted species may be obtained in reasonable time periods using far-infrared Fourier transform spectroscopy. They believe that the ratioing technique illustrated in Fig. 9.16c gives superior results to matrix Raman spectroscopy as far as relatively weak low frequency absorption bands are concerned.

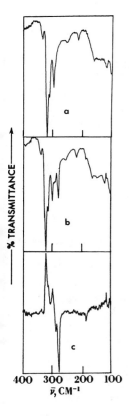

Fig. 9.16. (a) Far-infrared absorption spectrum of matrix isolated GeH_2Br_2 in an argon matrix at 20 K, ratioed against the cryostat with pure argon, measured using a Digilab FTS–16 Fourier transform spectrophotometer; (b) spectrum of the above sample after 5 hr *in situ* photolysis with a medium pressure mercury lamp; (c) spectrum resulting from ratioing trace (b) against (a). (Reproduced from [78] by permission of the author and the Society for Applied Spectroscopy; copyright © 1973.)

REFERENCES

1. EUCHEM Conference on Far Infrared Spectroscopy, Culham, England (1965).
2. MISFITS (Mellon Institute Symposium on Far Infrared Transpose Spectroscopy) Pittsburgh (1965).
3. A. Finch, P. N. Gates, K. Radcliffe, F. N. Dickson, and F. F. Bentley, *Chemical Applications of Far Infrared Spectroscopy*, Academic Press, New York, 1970.
4. K. D. Möller and W. G. Rothschild, *Far Infrared Spectroscopy*, Wiley-Interscience, New York, 1971.
5. J. R. Ferraro, *Anal. Chem.*, **40**(4), 24A (1968).
6. R. A. Nyquist and R. O. Kagel, *Infrared Spectra of Inoganic Compounds*, Academic Press, New York, 1971.
7. D. M. Adams and H. A. Gebbie, *Spectrochim. Acta*, **19**, 925 (1963).
8. M. Goldstein, E. F. Mooney, A. Anderson, and H. A. Gebbie, *Spectrochim. Acta*, **21**, 105 (1965).
9. M. J. Campbell, R. Grzeskowiak, and M. Goldstein, *Spectrochim. Acta*, **24A**, 1149 (1968).

10. M. Goldstein and W. D. Unsworth, *Inorg. Chim. Acta*, **4**, 342 (1970).
11. M. Goldstein and W. D. Unsworth, *Spectrochim. Acta*, **28A**, 1297 (1972).
12. K. S. Pitzer and S. J. Strickler, *J. Chem. Phys.*, **41**, 730 (1964).
13. W. H. Smith and G. E. Leroi, *J. Chem. Phys.*, **45**, 1767 (1966).
14. W. H. Smith and G. E. Leroi, *J. Chem. Phys.*, **45**, 1778 (1966).
15. P. R. Griffiths and H. W. Thompson, *Proc. Roy. Soc. (Lond.)* **A298**, 51 (1967).
16. S. M. Craven and F. F. Bentley, *Appl. Spectrosc.*, **26**, 242 (1972).
17. S. M. Craven and F. F. Bentley, *Appl. Spectrosc.*, **26**, 449 (1972).
18. S. M. Craven and F. F. Bentley, *Appl. Spectrosc.*, **26**, 484 (1972).
19. S. M. Craven and F. F. Bentley, *Appl. Spectrosc.*, **26**, 646 (1972).
20. W. G. Fateley and F. A. Miller, *Spectrochim. Acta*, **17**, 857 (1961).
21. W. G. Fateley and F. A. Miller, *Spectrochim. Acta*, **18**, 977 (1962).
22. W. G. Fateley and F. A. Miller, *Spectrochim. Acta*, **19**, 611 (1963).
23. E. C. Tuazon and W. G. Fateley, *J. Chem. Phys.*, **54**, 4450 (1971).
24. H. C. Silver and J. L. Wood; Trans. Faraday Soc., **60**, 5 (1964).
25. F. A. Miller, W. G. Fateley, and R. E. Witkowski, *Spectrochim. Acta*, **28A**, 894 (1967).
26. G. L. Carlson and W. G. Fateley, *J. Phys. Chem.*, **77**, 1157 (1973).
27. W. G. Fateley, *Pure and Applied Chem.*, **36**, 109 (1973).
28. G. L. Carlson, W. G. Fateley, and F. F. Bentley, *Spectrochim. Acta*, **28A**, 177 (1972).
29. L. Radom, W. J. Hehre, J. A. Pople, G. L. Carlson, and W. G. Fateley, *Chem. Comms.*, p. 308 (1972).
30. W. J. Lafferty, D. W. Robinson, R. V. St. Louis, J. W. Russell, and H. L. Strauss, *J. Chem. Phys.*, **42**, 2915 (1965).
31. J. A. Greenhouse and H. L. Strauss, *J. Chem. Phys.*, **50**, 124 (1969).
32. J. R. Durig and R. C. Lord, *J. Chem. Phys.*, **45**, 61 (1966).
33. T. R. Borgers and H. L. Strauss, *J. Chem. Phys.*, **45**, 947 (1966).
34. R. B. Sanderson and H. E. Scott, Aspen Int. Conf. on Fourier Spectrosc., 1970, G. A. Vanasse, A. T. Stair and D. J. Baker, Eds., AFCRL–71–0019, p. 167.
35. R. T. Hall, D. Vrabec, and J. M. Dowling, *Appl. Opt.*, **5**, 1147 (1966).
36. J. M. Dowling and R. T. Hall, *J. Mol. Spectrosc.*, **19**, 108 (1966).
37. R. T. Hall and J. M. Dowling, *J. Chem. Phys.*, **45**, 1899 (1966).
38. P. R. Griffiths and H. W. Thompson, *Spectrochim. Acta*, **24A**, 1325 (1968).
39. H. A. Gebbie et al., *J. Mol. Spectrosc.*, **19**, 7 (1966).
40. H. A. Gebbie, G. Topping, R. Illsley, and D. M. Dennison, *J. Mol. Spectrosc.*, **11**, 229 (1963).
41. J. E. Harries, N. R. W. Swan, J. E. Beckman, and P. A. R. Ade, *Nature*, **236**, 159 (1972).
42. W. J. Burroughs, E. C. Pyatt, and H. A. Gebbie, *Nature*, **212**, 387 (1966).
43. J. A. A. Ketelaar, J. P. Colpa, and F. N. Hooge, *J. Chem. Phys.*, **23**, 413 (1955).
44. Z. J. Kiss, H. P. Gush, and H. L. Welsh, *Can. J. Phys.*, **37**, 362 (1959).
45. Z. J. Kiss and H. L. Welsh, *Can. J. Phys.*, **37**, 1249 (1959).
46. D. R. Bosomworth and H. P. Gush, *Can. J. Phys.*, **43**, 729 (1965).
47. D. R. Bosomworth and H. P. Gush, *Can. J. Phys.*, **43**, 751 (1965).
48. H. A. Gebbie, N. W. B. Stone, and D. Williams, *Mol. Phys.*, **6**, 215 (1963).
49. R. J. Jakobsen and J. W. Brasch, *J. Am. Chem. Soc.*, **86**, 3571 (1964).

50. G. W. Chantry and H. A. Gebbie, *Nature*, **208**, 378 (1965).
51. S. G. Kroon and J. van der Elsken, *Chem. Phys. Letters*, **1**, 285 (1967).
52. P. R. Griffiths, unpublished work (1966).
53. B. J. Bulkin, *Helv. Chem. Acta*, **52**, 1348 (1969).
54. J. W. Fleming, P. A. Turner, and G. W. Chantry, *Mol. Phys.*, **19**, 853 (1970).
55. M. Davies, G. W. F. Pardoe, J. Chamberlain and H. A. Gebbie, *Trans. Faraday Soc.*, **66**, 273 (1970).
56. G. J. Davies and J. Chamberlain, *J. Phys.*, **A5**, 767 (1972).
57. R. F. Lake and H. W. Thompson, *Proc. Roy. Soc.*, (*Lond.*) **A291**, 469 (1966).
58. R. K. Thomas and H. W. Thompson, *Proc. Roy. Soc.* (*Lond.*), **A316**, 303 (1970).
59. R. K. Thomas, *Proc. Roy. Soc.* (*Lond.*), **A322**, 137 (1971).
60. R. K. Thomas, *Proc. Roy. Soc.* (*Lond.*), **A325**, 133 (1971).
61. H. A. Gebbie, W. J. Burroughs, J. Chamberlain, J. E. Harries, and R. G. Jones, *Nature* **221**, 143 (1969).
62. A. A. Viktorova and S. A. Zhevakin, *Sov. Phys. Dokl.*, **11**, 1059 (1967).
63. H. A. Gebbie, W. J. Burroughs, J. E. Harries, and R. M. Cameron, *Astrophys. J.*, **154**, 405 (1968).
64. S. G. W. Ginn and J. L. Wood, *Trans. Faraday Soc.*, **62**, 777 (1966).
65. R. F. Lake and H. W. Thompson, *Spectrochim. Acta*, **24A**, 1321 (1968).
66. J. Yarwood and W. B. Person, *J. Am. Chem. Soc.*, **90**, 594 (1968).
67. A. Anderson, S. H. Walmsley, and H. A. Gebbie, *Phil. Mag.*, **7**, 1243 (1962).
68. A. Anderson, H. A. Gebbie, and S. H. Walmsley, *Mol. Phys.*, **7**, 401 (1964).
69. S. H. Walmsley and A. Anderson, *Mol. Phys.*, **7**, 411 (1964).
70. A. Anderson and S. H. Walmsley, *Mol. Phys.*, **9**, 1 (1965).
71. A. Anderson and S. H. Walmsley, *Mol. Phys.*, **7**, 583 (1964).
72. A. Anderson and S. H. Walmsley, *Mol. Phys.*, **10**, 391 (1966).
73. A. Anderson and W. H. Smith, *J. Chem. Phys.*, **44**, 4216 (1966).
74. S. S. Mitra, *Solid State Phys.*, **13**, 1 (1962).
75. D. P. Martin, *Adv. Phys.*, **14**, 30 (1965).
76. J. W. Brasch and R. J. Jakobsen, *Spectrochim. Acta*, **20**, 1644 (1964).
77. A. J. Barnes et al., *Chem. Comms.*, p. 1089 (1969).
78. W. A. Guillory and G. R. Smith, *Appl. Spectrosc.*, **27**, 137 (1973).

MID-INFRARED ABSORPTION SPECTROSCOPY

I. "CONVENTIONAL" INFRARED SPECTROSCOPY

The type of samples most commonly encountered by practicing infrared spectroscopists are KBr discs, mulls, solutions, and capillary films; the first question that is asked about Fourier transform spectroscopy by spectroscopists used to handling these conventional samples often concerns any differences between spectra measured on a grating spectrometer and spectra measured interferometrically. Even though there are a few differences between spectra measured on each type of instrument, they are usually so small that their effects are seldom apparent. For example, many spectra measured on good grating spectrometers are run using slit programs designed to give constant energy at the detector for all frequencies in the range so that the resolution of grating spectra is generally lower at the ends of the spectrum than at the center. This effect is not observed in spectra measured using a Fourier transform spectrometer, for which a constant resolution is found across the entire spectrum. However, most samples whose infrared spectrum is of interest in the typical analytical laboratory do not have narrow enough bands for this effect to be apparent.

Similarly, if triangular apodization is used in the computation of the spectrum from an interferogram, there will be little or no difference in the line shape of absorption bands measured using grating or Fourier transform spectrometers. If the interferogram is not apodized, side-lobes will, of course, be seen on bands whose half-width is less than the resolution. Thus, for most spectra encountered by analytical infrared spectroscopists, there are no obvious differences between grating and interferometric spectral data unless the spectra are noisy. In view of the nature of the sampling process, the period of the noise in spectra measured by Fourier transform spectrometers is always the same as the resolution; for spectra measured using a grating monochromator, it is customary to set the time constant of the amplifier and the pen response time such that the period of the noise is less than the resolution in order to avoid dynamic error [1].

These factors do not seriously affect spectra of the type now measured using prism or grating spectrometers, since the noise level on interferometrically measured spectra of samples of this type is usually extremely low. In fact in a recent paper, Smith and Potts [2] suggested that "Class II" reference spectra [3] can easily be measured by contemporary Fourier transform spectrometers.

There are many reasons for measuring spectra under carefully controlled instrumental conditions, and the Coblentz Society has established a set of criteria for the classification of infrared reference spectra for the purpose of allowing spectra measured on one instrument to be validly compared against the spectrum of the same sample measured using a different instrument. A Class I spectrum was defined as a spectrum completely independent of the spectrometer. A Class II spectrum, or research quality analytical spectrum, was defined as the spectrum of a pure material measured on a good grating spectrometer "operated at maximum efficiency under conditions consistent with accepted laboratory practice". Class III spectra, or approved analytical reference spectra, are those produced on defined substances, again using good sampling techniques, with a high quality NaCl prism spectrometer or a grating spectrometer not meeting Class II standards.

At the time that the standards for each of these classes of infrared spectra were first set down, no commercially available Fourier transform spectrometer was capable of meeting the specifications required even for the measurement of Class III spectra. Now that high-quality, high-resolution infrared Fourier transform spectrometers are commercially available, it is apparent that they should meet at least the criteria of the instruments which are approved for the measurement of Class II spectra. If not, their use in many analytical chemistry laboratories will be severely limited.

The specifications for operating a grating spectrometer for the measurement of Class II spectra are as follows.

Frequency range: The spectrum should cover at least the range from 3800 to 450 cm^{-1}.

Resolution: The spectral slit width should not exceed 2 cm^{-1} through at least 80% of the wave number range.

Wavenumber accuracy: Wavenumbers as read from the chart should be accurate to ± 5 cm^{-1} above 2000 cm^{-1} and ± 3 cm^{-1} below 2000 cm^{-1}.

Noise level: The noise level must not exceed 1% average peak-to-peak (or 0.25% R.M.S.).

Scattered light: Apparent stray radiation should be less than 1% above 500 cm^{-1}.

I_o line flatness: With no cell in either beam the I_o line (100% line) should be flat to ± 0.01 absorbance units when the spectrometer is used under the same conditions used for the measurement of the reference spectrum.

It is of interest to see if modern high resolution mid-infrared Fourier transform spectrometers can meet all these criteria. Laser fringe-referenced interferometers can certainly meet the frequency accuracy specifications. The frequency range criterion is just able to be met; the range of 3800 to 450 cm^{-1} represents the maximum useful range of a germanium beamsplitter on a KBr substrate. The resolution of a Fourier transform spectrometer is constant across the spectrum and 2 cm^{-1} resolution is easily achieved with most interferometers; with a retardation of 0.5 cm the apodized spectrum has a nominal resolution of about 3 cm^{-1} but the same interferogram computed with boxcar truncation yields a spectrum whose resolution is approximately 1.5 cm^{-1}. Base-line flatness is equally easy to achieve, especially if the technique of using the same beam for the sample and reference spectra is used. "Dynamic error" is a problem which can be found with grating spectrometers if the gain or pen response is set too low [1], but no similar problem exists with interferometers.

Therefore, just two specifications, those for stray light and noise level, are the only ones that need to be met and these require careful consideration. If efficient optical and electrical filtering of the spectrum above the folding frequency is carried out, no stray light will be seen in the high frequency region of the spectrum. However, many optical filters with sharp high frequency cut-offs do not have good low frequency transmission. For this reason optical filters are often left out and solely electronic filtering is used. Since a sharp cut-off is difficult to achieve with band-pass filters there is a chance of folding high frequency radiation back in the region below 3800 cm^{-1} unless a short sampling interval is used.

It seems surprizing that noise level may be a problem for Fourier transform spectrometers measuring the light from intense incandescent sources, and indeed the noise level in the central region of the spectrum will be very low even with very short scan times. The main problem exists at the extremes of the spectral range where the beamsplitter efficiency falls off. Even though a noise level well under 1% can be found around 2000 cm^{-1} in a very short measurement time, between 3800 and 3600 cm^{-1} and 600 and 450 cm^{-1} a noise level which is considerably in excess of this value may be found. Unless double-precision methods are applied when data systems with 16-bit words are used for data collection, it may not be possible to reduce the noise-level at the extremes of the spectrum even after extended signal averaging.

In practice, it has been found in a survey [2] of the performance of several grating spectrometers and one Fourier transform spectrometer (the FTS–14) that Class II spectra are able to be measured using the Fourier transform spectrometer. No Class I spectra have yet been measured on any spectrometer, and yet in the same article which compared the performance of spectrometers for the measurement of Class II reference spectra, Smith and Potts have surmised that Fourier transform spectrometers may make such spectra a real possibility within the next few years.

At the present time, only those laboratories where many Class II standard reference spectra are measured daily can justify the use of a Fourier transform spectrometer for routine spectroscopy. Most owners of these instruments use them for solving problems that would be particularly time consuming if grating spectrometers were used. Few of these problems involve the measurement of the absorption spectra of the type of samples commonly encountered in infrared spectroscopy such as Nujol mulls, 13 mm KBr discs, films, and solutions.

A paper by Low and Freeman [4] does describe one application where a Fourier transform spectrometer was used for the examination of weak features in the absorption spectra of a series of natural oils. These authors were able to demonstrate significant differences in the spectra of orange oils from different sources by scale expanding several regions of the spectrum between 1830 and 900 cm^{-1} (Fig. 10.1). These orange oils contain about 95% of the terpene hydrocarbon limonene, the remaining few percent comprising more than 100 constituents. Analysis of these constituents by gas-liquid chromatography can be very time consuming but at present it is the preferred technique. Analysis of the infrared spectrum using Fourier transform spectroscopy certainly shows differences between these samples and yet the cause of each new spectral feature is difficult to assign due to the predominance of the bands of limonene. The application of a program for subtracting the absorbance spectrum of pure limonene to leave only the absorption features of the minor components will be necessary before Fourier transform spectroscopy can be truly applicable to this problem.

Another study related to the investigation described above involved the measurement of the spectrum of an antioxidant present at very low concentration in a polymer sheet [5]. In this project a large thickness of the sheet was required in order to detect any band of the antioxidant, so that most of the spectrum was "blacked out." The actual sensitivity for the determination of trace impurities using contemporary equipment naturally depends on the intensity of the spectral structure of the matrix material; in favorable cases trace components in the low parts-per-million concentration range may be studied. The highest sensitivity for any determination yet achieved by Fourier transform spectrometer is in the determination of oxygen present in high quality silicon [6]. Here a sensitivity of less than 10 ppb has been achieved.

When the spectrum of a sample in solution is to be measured, it is always desirable to eliminate any absorption due to the solvent. The problems involved in measuring the spectra of solutions are essentially the same for Fourier transform spectroscopy as for grating spectrometers. The best results are always obtained when the solvent has no absorption bands in the spectral range of interest while the bands due to the solute are strong enough that they require no ordinate scale expansion. When the solvent shows no absorption bands in the frequency range of interest, exact matching of the sample and reference cells is not needed. However, this situation is unusual in infrared

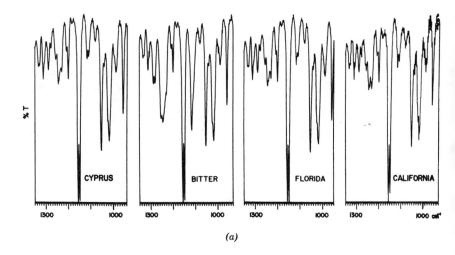

CYPRUS   BITTER   FLORIDA   CALIFORNIA

(a)

BITTER   CALIFORNIA   CYPRUS   FLORIDA

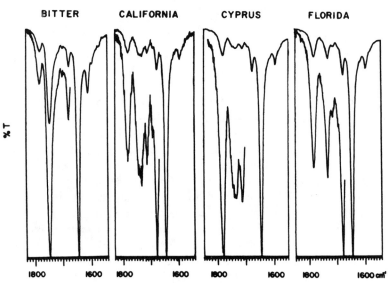

Fig. 10.1. Ordinate scale-expanded spectra of various orange oils, as capillary films, (a) between 1350 and 950 cm^{-1} and (b) between 1830 and 1540 cm^{-1}. Minor differences in composition between oils from different sources are indicated. (Reproduced from [4] by permission of the author and the American Chemical Society; copyright ©️ 1970.)

248

spectroscopy; even when solvents with very few absorption bands are used, at least two solvents (e.g., CCl_4 and CS_2) have to be used to adequately cover the mid infrared spectrum from 3800 to 450 cm^{-1}.

In practice it is often quite easy to measure the spectra of dilute solutions while compensating for the solvent using Fourier transform spectrometers with a good data system since the single-beam spectrum of the pure solvent may be measured and stored in the computer memory. The same cell used to hold the pure solvent may then be used to hold the solution so that the path lengths of the cells for the sample and reference spectra are identical. This technique will only eliminate solvent bands for very dilute solutions. For more concentrated solutions, the solute displaces a certain amount of the solvent from the volume contained in the cell so that positive-going peaks due to the solvent are seen in the ratio-recorded spectrum unless the solvent bands are compensated digitally in the way described in Chapter 6, Section IV.

In the measurement of solution spectra, solvent bands may often transmit only a very small proportion, say 1%, of the incident energy in regions where there are strong absorption bands. Although a 1% transmittance is usually sufficient for a transmittance spectrum to be measurable across the solvent bands, the noise level in the ratio-recorded spectrum may be higher than expected. Besides resulting from the attenuation of the reference signal across the absorption bands of the solvent, excess noise may also result from the limited dynamic range of the signal-averaging computer (for a rapid-scanning interferometer) or of the analog-to-digital converter (for both rapid- and slow-scanning interferometers). For most solvents, the majority of the radiation reaching the detector, and therefore contributing to the large signal at the zero retardation point, is transmitted in the "window" regions *between* the absorption bands, and the proportion of this signal due to frequency regions over which the liquid has significant absorption can be quite small. Therefore, unless both data collection and the FFT are performed in double precision with a rapid-scanning interferometer, the signal-to-noise ratio may never be raised above a certain value after the dynamic range of the interfero-gram reaches the word-length of the signal-averaging computer. In situations such as this one, a rapid-scanning interferometer will always give superior results compared with a slow-scanning interferometer, since the dynamic range of a *signal-averaged* interferogram can, in theory, be raised to any desired amount.

This effect is particularly noticeable when the spectra of aqueous solutions are measured. The spectrum of water shows very strong absorption bands at about 3400 and 1640 cm^{-1} and a weaker band at 2120 cm^{-1}. For path-lengths greater than about 15 μm, the transmittance of water is less than 1% at 1640 cm^{-1}, so that great care has to be taken when measuring the spectra of aqueous solutions in the region of this band. As it happens many materials whose spectra are of interest when the sample is in aqueous solution have absorption bands close to 1640 cm^{-1}; for example, proteins have been examined in aqueous solution, and the important amide I band is usually observed only a few wavenumbers from the center of the water band. Between about 1750 and 2900 cm^{-1}, the transmittance of a 15 μm film of water is quite high so that it does not take much signal-averaging before the dynamic range of the interferogram exceeds 2^{16}, even when a TGS detector is used. For this reason interferograms of these samples must always be collected and computed in double-precision if the data system has a word-length of 16 bits.

Results of importance to soil scientists have recently been published by McCarthy et al. [7], in which the spectra of humic materials have been measured in aqueous solution using a rapid-scanning Fourier transform spectrometer. The samples are very complex mixtures of organic compounds containing carboxylic acid and phenolic groups. Their spectra are strongly pH dependent and also change when certain metal ions are added to the solution due to the complexing action of the metal ions; studies of the spectra of these materials may give new information on the mechanism by which metabolically important metals are transported through soil. Until the advent of FTS, the spectra of humic materials could only be measured with the sample prepared as a Nujol mull, as a KBr disc, cast on a AgCl plate or as a solution in D_2O. Whereas the use of D_2O more closely resembles the natural environment than any of the other techniques, exchange of D_2O and OH groups on the humic acid causes some problems of interpretation of the spectra, and spectra measured with the sample in H_2O are far preferable. The spectra of humic acid in H_2O are shown in Fig. 10.2; it can be seen that although the noise level is higher around 1640 cm^{-1} than in other spectral regions, the shape of the spectra in this region is still very apparent.

Koenig and his co-workers [8] have recently measured the spectra of several proteins in aqueous solution at path-lengths of between 5 and 10 μm, using a Fourier transform spectrometer. At these short path-lengths, the problem of noise around 1640 cm^{-1} is not so critical as the previous measurements to be described, but it was found that such concentrated solutions had to be used that the spectra always had to be digitally corrected to accurately compensate for the water band. Typical spectra measured in this way are shown in Fig. 10.3.

The effects of digitization were particularly critical in a series of measurements carried out by Alben [9] who needed to quantitatively analyze a series of aqueous samples containing a functional group which absorbs at approximately 2120 cm^{-1}, the same frequency as a weak band in the water spectrum. Since the samples were only slightly soluble in water, it was found to be necessary to increase the thickness of the cell to obtain measurably strong absorption bands due to the solute. At the point when the band of interest

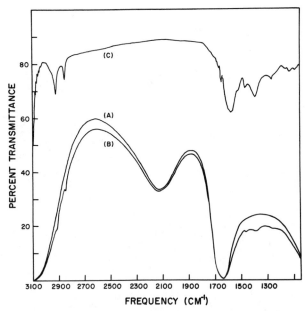

Fig. 10.2. (a) Transmittance spectrum of a 17.6 μm thick film of water; (b) transmittance spectrum of the same thickness of a solution of humic acid in NaOH; (c) the result of ratioing spectrum A versus spectrum B; note how much sharper the bands around 2900 cm^{-1} become when the effect of the strong solvent absorption is removed, and also note the fact that it is possible to measure absorption features under the very strong water band at 1640 cm^{-1}. (Reproduced from [7] by permission of the American Chemical Society; copyright © 1975.)

absorbed 1%, the water solvent only *transmitted* 1% at 2120 cm^{-1}, even though the regions on either side showed good transmittance (Fig. 10.4a). To attain the maximum sensitivity for this determination, a liquid nitrogen cooled InSb detector was used; however the energy transmitted on either side of 2120 cm^{-1} was so large that the dynamic range of the ADC was then exceeded. In order to allow interferograms to be signal-averaged to increase the S/N, a narrow band-pass optical filter which only transmitted radiation around 2100 cm^{-1} was inserted in the beam so that large variations in the intensity of the reference spectrum were no longer observed and detector noise became larger than digitization noise, as shown in Fig. 10.4b. Using this arrangement the bands of interest could be measured at a sufficiently high signal-to-noise ratio that quantitative measurements could be performed.

The obvious disadvantage of this method is that Fellgett's advantage is

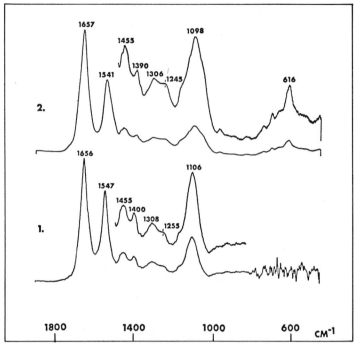

Fig. 10.3.   Infrared absorbance spectra of hemoglobin: (1) Aqueous solution, pH = 4.8, (2) Cast film on AgCl plate. The change in frequencies of the bands associated with the amide group has been assigned to differences in the hydrogen bonding pattern of the proteins in solution and as a cast film. (Reproduced from [8] by permission of the author.) and Wiley-Interscience; copyright ⓒ 1975.)

decreased owing to the reduction of the frequency range able to be studied because of the optical filter. Thus for this type of experiment, the use of a Fourier transform spectrometer is less beneficial than in experiments where the complete spectrum is of interest. In fact, if in any experiment only one band is to be studied at low resolution, the use of a Fourier transform spectrometer will rarely give better results than a grating spectrometer with the same type of source and detector.

II. STRONGLY SCATTERING OR ABSORBING SAMPLES

One particular type of sample for which infrared spectroscopy has not been able to be applied routinely in the past scatters or absorbs the radiation from the source to such an extent that only a small proportion reaches the detector.

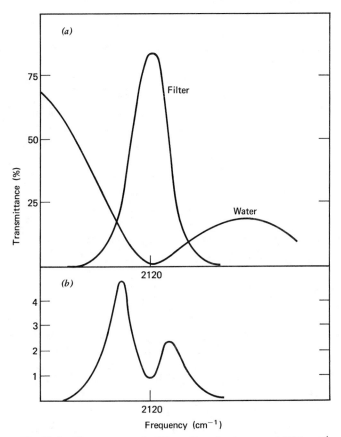

Fig. 10.4. The spectrum of a 100-μm film of water around 2120 cm^{-1}, showing a minimum transmittance of about 1%. The high transmittance on either side of this band makes the dynamic range of the ADC the limiting source of noise when a sensitive detector is used for the measurement. By filtering out information from unwanted spectral regions with a narrow band-pass filter, the overall transmittance of the water and filter has the profile shown in (b), so that the spectrum once more becomes detector noise limited.

The amount of scattering increases with frequency so that spectra at high frequency are particularly difficult to measure. The types of sample which fall into this category include paper, certain types of cloth, blackened rubber, and poorly ground or compressed powders.

One of the more significant studies in which scattering samples have been investigated using Fourier transform spectroscopy was performed by Low, Goodsel, and Takezawa [10, 11]. They describe an infrared study of the sorption of SO_2 on CaO [10] and MgO [11]. In this investigation thin discs of CaO and MgO of 20 cm diameter were prepared by pressing approximately 65 mg of the powder in a steel die at a pressure of 70–80 tons per square inch. Clear discs have not been able to be produced under these conditions, neither are they wanted since the surface area of the sample should be as large as possible to achieve the maximum amount of sorption. The N_2 BET surface area of the samples prepared in this way after degassing was 10.7 m^2/gm for CaO and 16.5 m^2/gm for MgO. The nature of these samples is such that a great deal of scattering and absorption occurs and great demands are made on the spectrometer in order to obtain spectra of adequate resolution and signal-to-noise ratio. Low [12] has stated that similar spectra could be recorded in periods ranging from one to three hours with a Perkin-Elmer Model 621 grating spectrometer although their resolution was lower and good definition of the bands was not achieved below 700 cm^{-1}. The FTS–14 spectrometer used for this study produced good spectra in 5 to 10 min. This was not only helpful from the point of view of saving time but also because the interactions being studied by Low et al. [10, 11] were time-dependent and spectra had to be measured as quickly as possible.

Low and his co-workers were able to deduce mechanisms for the sorption and subsequent chemical reactions of SO_2 on the surface of the oxides. Their work will be illustrated from the calcium oxide study. As SO_2 is added to the sample, a strong doublet at 973 and 925 cm^{-1} is seen in the spectrum, which increases in strength and gradually merges into a singlet as more SO_2 is added. Low was able to assign this doublet to formation of the sulfite ion. Two other bands at 1344 cm^{-1} and 1150 cm^{-1} are seen which are apparently due to physically adsorbed SO_2 and disappear rapidly when the sample is subjected to a vacuum. On degassing at temperatures above 500°C marked changes occurred in the spectrum due to decomposition of the sulfite. The spectra produced are complex (Fig. 10.5a and b), with too much structure for a simple mechanism to be assigned for the surface reactions. Low et al. suggest that the calcium sulfite disproportionates as:

$$4 \, CaSO_3 \rightarrow 3 \, CaSO_4 + CaS$$

with the high temperature allowing sulfite ions to migrate to suitable reaction sites. The infrared bands produced at higher temperatures are assigned to species of the general form $S_x O_y^{n-}$ resulting from the polymerization of undecomposed SO_3^{2-}, and sulfate and sulfide ions produced by disproportionation. These spectra require fairly high resolution for all the bands present to be resolved; however since no differences were noticed on changing from 2 cm^{-1} to 1 cm^{-1} resolution, the spectra may be considered to be free of any instrumental effects.

In another type of measurement where scattering effects have prevented good results being obtained in the past, Percival and Griffiths [13] developed

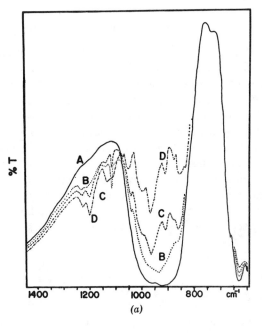

(a)

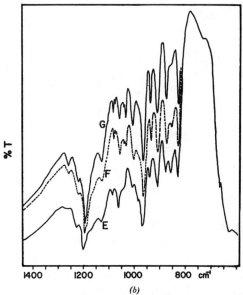

(b)

Fig. 10.5. Spectra of SO_2 adsorbed on CaO at 25°C and measured after evacuation for 18 hr (spectrum A). The sample was then degassed sequentially for 3 hr at the following temperatures: B, 550°C; C, 560°C; D, 580°C; E, 650°C; F, 665°C; G, 750°C. (Reproduced from [10] by permission of the author and the American Chemical Society; copyright © 1971.)

a method for identifying components separated by thin-layer chromato-graphy (TLC) directly on the plate. Thin layers (~ 100 μm) of silica gel or alumina were prepared, not on conventional TLC substrates such as glass, aluminum or polyethylene sheets, but rather on an infrared transmitting substrate which is not attacked by most solvents used in TLC, silver chloride. By ratioing the single-beam spectrum of the plate in a position in which the sample is known to be present to that of a nearby region where no sample is present, transmittance spectra of samples free from any features due to the substrate can be measured. The sensitivity of this technique is such that micro-gram quantities of materials can readily be measured; for example, Fig. 10.6 shows the spectrum of a one microgram spot of a dye, Sudan Red G, measured in this way after separation from a mixture of other dyes.

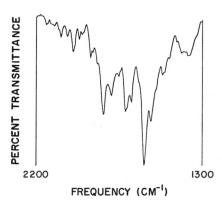

Fig. 10.6. The spectrum of 1 μg of Sudan Red G measured by direct transmittance through a TLC plate consisting of a 100-μm layer of silica gel deposited on a sheet of silver chloride. The dye was originally present as a 0.1% component in a 1-μl spot of a solution containing three dyes, Sudan Red G, Indophenol Blue, and Butter Yellow. (Repro-duced from [13] by permission of the Amer-ican Chemical Society; copyright © 1974.)

Carbon-filled rubber samples have long presented difficulty for analytical spectroscopists due to the strong absorption of infrared radiation by carbon black. Koenig and Tabb [14] have compared the spectra obtained from the same KBr pellet of carbon-filled rubber measured on an optical-null grating spectrometer and a rapid-scanning Fourier transform spectrometer; their spectra are shown in Fig. 10.7. The poor response of the servo-mechanism of the grating spectrometer when the energy in the reference beam is strongly attenuated is very obvious from these spectra. Frequency shifts of the C=C double bond stretching mode of the rubber samples on adding carbon black have been measured by Koenig and Tabb, who suggest that this shift is caused by complexing between the carbon black and rubber. This complexing ap-parently gives rise to the well-known stiffening effects of carbon black.

In measurements on another type of poorly transmitting sample—cloth—the effect of scattering was surprizingly found to be very small [5]. Ratio-recorded spectra of cloth samples showed that the proportion of energy from

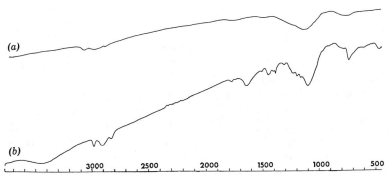

Fig. 10.7. The transmittance spectra of a film of carbon-filled rubber, (a) as measured on a grating spectrophotometer, and (b) as measured on a Fourier transform spectrometer. (Reproduced from [14] by permission of the author and McLean-Hunter Ltd.; copyright © 1974.)

the source that reached the detector did *not* decrease with increasing frequency as would be expected if scattering were taking place to any great extent; rather the base-line was quite flat across the entire spectrum. The fact that fairly strong absorption bands due to surface treatment of cotton fibers were observed in the spectrum suggested that the radiation reaching the detector was not transmitted through small holes in the weave but apparently was transmitted through the outer edges of the fibers without causing any scattering. This observation might lead to a powerful method of studying thin surface coatings on fibers.

From the results described in this section, it is obvious that Fourier transform spectroscopy provides the analytical chemist with a powerful new tool for the study of samples which strongly attenuate the intensity of transmitted radiation. Measurements have been described where the cause of this attenuation is scattering (in the surface studies of Low et al. and the TLC measurements of Percival and Griffiths), general absorption (the measurement of carbon-filled rubber samples) or vignetting (the measurements on the cloth samples). It may be easily forecast that many more results of this type will be announced in the future.

III. SPECTROSCOPY OF MICROSAMPLES

In view of the good performance of Fourier transform spectroscopy for poorly transmitting samples, FTS should be applicable for microsampling where the area of the sample is so small that only a small proportion of the incident energy is usually transmitted through the cell and a large fraction is cut out due to the small aperture. To envisage the energy at the detector of a

Fourier transform spectrometer, let us consider how this energy is reduced as the area of the sample is decreased. If the normal diameter of the beam in the sample compartment is 12 mm and a 4x beam-condenser is used to support the sample, 3 mm diameter samples may be examined with only the attenuation due to the transmission of the beam-condenser. The diameter of microdiscs commonly used in infrared microsampling is 1.5 mm or 0.5 mm. The energy at the detector is reduced by the ratio of the area of the aperture of the cell to the area of the focus in the beam-condenser, and so for a 1.5 mm diameter cell the energy is reduced by a factor of four; for a 0.5 mm diameter sample the area is reduced by a further factor of nine.

As the size of the sample is reduced still further, the energy is reduced to the point where spectra are extremely difficult to measure by conventional spectrometers and yet can still be fairly easily measured using interferometry. King has described one application of the measurement of small particles by Fourier transform spectroscopy [15]. A black contaminant, 250 μm in diameter, was found in a molded plastic plate; the plastic was removed and the spectrum of the spot was measured directly. A spectrum of a portion of a dead insect found in the bin of the granulated polymer feed stock was also measured in the same way. The similarity of the two spectra (Fig. 10.8) indicated the source of the contamination of the plastic plate.

King has stated that the analysis of particles as small as 100 μm in diameter can be performed, and he has measured the spectrum of a sheet of polyethylene through an aperture 50 μm in diameter using a TGS detector. This work was done using a very well-aligned 6x beam-condenser so that little energy was lost when no sample was present in the beam, giving an increase in sensitivity of approximately four times that for a 4x beam-condenser used with a 50% energy loss.

In independent studies, King [15] and Griffiths and Block [16] both estimate a theoretical detection limit of approximately one nanogram for a Fourier transform spectrometer used under optimum conditions with a TGS detector. This figure is a theoretical value and was arrived at by masking down a sample of known concentration until the bands could only just be seen in the spectrum. This procedure does not yield a very practical method of analyzing microsamples since it actually uses a sample containing much more of the material under study than is present in the infrared beam. It is more realistic to estimate the minimum total amount that can realistically be analyzed under laboratory conditions.

A standard technique for measuring the infrared spectrum of a fairly non-volatile material is to mix the powder with powdered KBr and press the material into a transparent disc. When the quantity of material to be analyzed is less than 10 micrograms, the material cannot be mixed directly with the KBr powder because the sample is too small to handle. Under these circum-

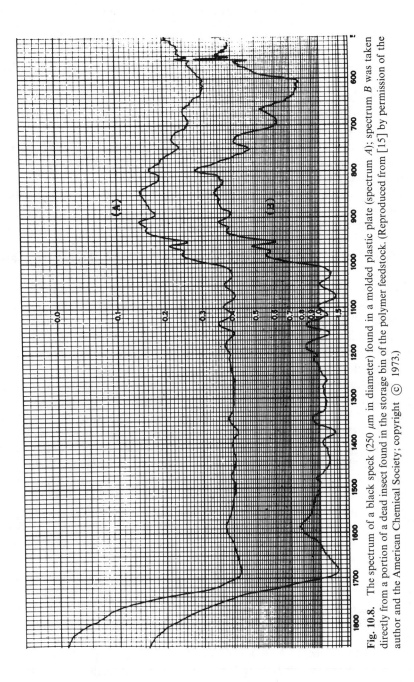

Fig. 10.8. The spectrum of a black speck (250 μm in diameter) found in a molded plastic plate (spectrum *A*); spectrum *B* was taken directly from a portion of a dead insect found in the storage bin of the polymer feedstock. (Reproduced from [15] by permission of the author and the American Chemical Society; copyright © 1973.)

stances the sample is generally dissolved in a volatile solvent, the solution is added to the KBr powder and the solvent is evaporated away. However unless the sample is extremely nonvolatile, the loss of sample can be very high during this evaporation step.

In a study of evaporation effects during microsampling, Griffiths and Block [16] used a series of phosphonate acid esters which were liquids of fairly low volatility, having a saturated vapor pressure of about 1 mm of Hg at 200°C and about 10^{-4} mm at 20°C. Assuming that the ideal gas laws hold, the saturated vapor containing 100 ng of these materials would occupy less than 150 ml at 20°C so that 10 ng would occupy only 15 ml, or less than one cubic inch. If this small amount were dissolved in a solvent and the solvent were evaporated off, it would only require the mixture to be in equilibrium with one cubic inch of air before 10 nanograms of the sample would have disappeared.

Even in the unlikely circumstance that not all of the sample evaporates during the evaporation of the solvent, more would be lost during the transfer of the mixture of KBr and sample to the die before pressing the pellet. To check this effect, a sufficiently large amount of a phosphonate acid ester was mixed directly with KBr so that if no loss of sample occurred while an aliquot was being transferred to the microdie a strong spectrum would have resulted. In fact the measured spectral bands were rather weak. On leaving several aliquots of this mixture exposed to the atmosphere at room temperature for periods up to half an hour, the spectral bands in the discs prepared from each aliquot became progressively weaker indicating that evaporation in the atmosphere was taking place. By plotting the absorbance of one particularly strong band in the spectrum of ethyl acid phosphonate against the time of exposure to the atmosphere, Griffiths and Block demonstrated that most of the sample loss occurred during the first few seconds of exposure (Fig. 10.9). It was shown that as much as 500 ng of these relatively nonvolatile materials could be lost during the transfer step before pressing a 0.5 mm diameter KBr microdisc. The use of a Fourier transform spectrometer was advantageous for this study since one spectrum could be measured in the short time that it took to prepare the next sample.

These conclusions have been verified by King [15] using 2,6-dimethoxyphenol, which is a solid at room temperature (M.Pt. = 55°C) and of lower volatility than the samples studied by Griffiths and Block. Using a Fourier transform spectrometer to enhance the sensitivity of his experiment, King demonstrated that if 0.5 μl of a 0.1% solution in CS_2 (0.5 μg of sample) is added to 0.2 mg of KBr and the solvent is evaporated away, sufficient sample remains in the KBr that a fairly strong spectrum can be obtained when the mixture is pressed into a 0.5 mm diameter disc. If the same amount

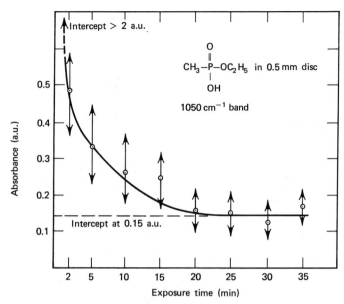

Fig. 10.9. Variation of the absorbance of one band in the spectrum of a rather involatile liquid, after aliquots of a mixture of the sample and KBr had been exposed to the atmosphere for increasing periods of time. If no evaporation had occurred, the absorbance from all of these samples would have been greater than 2 a.u. (Reproduced from [16] by permission of the Society for Applied Spectroscopy; copyright ©️ 1973.)

of sample is added to the same quantity of KBr, but using 5 μl of a 0.01% CS_2 solution, the amount of sample measured to be present in the disc decreases by a factor of four due to evaporation of the sample with the solvent. If 30 ng are present in the disc, several bands which are characteristic of 2,6-dimethoxyphenol can still be seen (Fig. 10.10). King's studied showed that for this compound most of the sample loss occurred while the solvent was being evaporated off and little loss occurred while the mixture with KBr was being transferred to the die.

King also studied the problem of transferring 1 to 2 μg of sample from a solution of between 1 and 20 ppm concentration using the Wick-Stick [17] method. Recovery was too low to obtain an identifiable spectrum in most cases even using a Fourier transform spectrometer. He showed that it was more efficient to concentrate a dilute solution into a few microliters using a microdistillation apparatus before the sample is transferred to the KBr. In no case using the micro KBr disc technique could King obtain the spectrum of less than 10 ng of any material.

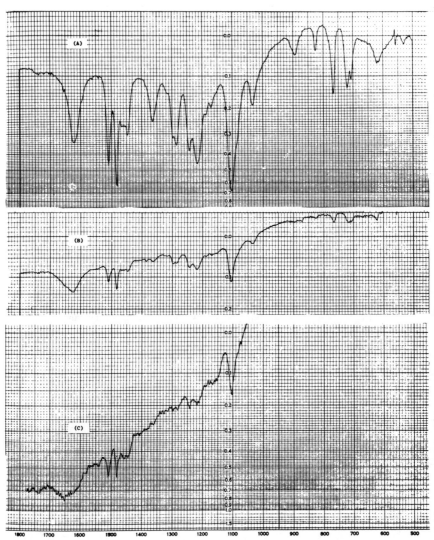

Fig. 10.10. The spectra of 2,6-dimethoxyphenol in 0.5-mm KBr microdiscs, measured on the FTS–14 spectrometer: (a) sample in 0.5 μl, 0.1% CS_2 solution, transferred to 0.2 mg of KBr, no scale expansion; (b) sample in 5.0 μl, 0.01% CS_2 solution, transferred to 0.2 mg of KBr, no scale expansion; (c) sample in 0.5 μl, 0.01% CS_2 solution, transferred to 0.2 mg of KBr, 7x ordinate scale expansion. (Reproduced from [15] by permission of the author and the American Chemical Society; copyright ⓒ 1973.)

262

The KBr disc technique has an additional disadvantage in that it is very difficult to prepare discs that have a uniform transmittance. When the baseline of a spectrum varies (either due to scattering, to minor impurities in the KBr or adsorbed water) it is extremely difficult to achieve large ordinate scale expansions over long spectral intervals in order to be able to observe very weak bands. It is much easier to obtain a flat baseline if the sample is held in a solution of, say, CS_2 or CCl_4 and the same thickness of the pure solvent is used for the reference spectrum. This method is ideally suited for Fourier transform spectroscopy. The only interference that prevents the maximum sensitivity being reached in this case is the difficulty of precisely cancelling out the effects of strong absorption bands of the solvent from the ratio-recorded spectrum. Using small cavity cells (1 mm diameter) the detection limit of Fourier transform spectroscopy for measuring the spectra of solutes is approximately equal to that of the micro KBr disc method, even though the area of the cell is four times larger than that of the 0.5 mm KBr micro-discs. For volatile materials, the solution method can increase the practical sensitivity by more than an order of magnitude.

The solution method is definitely superior when the components of very low concentration solutions are being studied. King evaporated 0.2 ml of a CS_2 solution of 2,6-dimethoxyphenol at a level of 2.5 ppm in a small centrifuge tube to about 5 μl volume, and recorded the spectrum of the resultant solution. He showed that 70–80% of the sample was recovered, which was a much higher recovery rate than that obtained by direct evaporation onto KBr powder. King has also stated that this technique is a good method of analyzing fractions separated by gas chromatography. The sample is trapped in a capillary tube and washed out with 3 μl of CS_2. He has also used solution spectroscopy for analyzing samples isolated on a TLC plate. If the sample occupies 1 cm^2 on a 0.1 mm thick TLC plate, the substrate can be scraped off and extracted with 50–200 μl of solvent. After concentrating this solution to a few microliters volume, the infrared spectrum is recorded. The success of this method depends on the volatility of the sample and the effectiveness of the solvent in removing the sample from the substrate. King applied 5 μg of 2,6-dimethoxyphenol to a 0.1 mm thick silica gel plate and developed with methylene chloride for 35 min. The silica gel at the sample spot (2.5 cm from the origin) was scraped off and extracted twice with 50 μl of CS_2. The total extract was evaporated down to 5 μl, placed in a 1 mm cavity cell and its spectrum was measured. Figure 10.11 shows the spectrum with and without scale expansion. About 40 ng of sample were estimated to have been recovered, the remainder having been lost largely through evaporation on the TLC plate and the inefficiency of transferring the sample by CS_2.

It can be seen from these examples of infrared microsampling, all of which were performed using a Fourier transform spectrometer, that in many cases

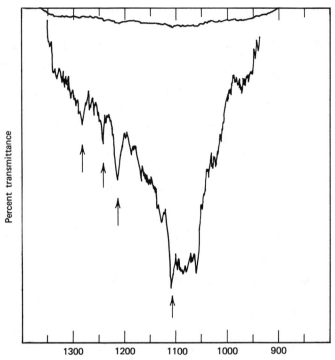

Fig. 10.11. The spectrum of 2,6-dimethoxyphenol in CS_2 solution after solvent absorptions were compensated by ratioing against an equal thickness (1 mm) of pure solvent. 5 μg of sample on a silica gel TLC plate were developed with CH_2Cl_2 for 35 min, after which time the spot had moved about 2.5 cm. The sample was extracted with 50 μl of CS_2 twice and the solution was evaporated to 5 μl. Upper trace, 1000 scans, no scale expansion; lower trace, same spectrum, 30x ordinate scale expansion. (Reproduced from [15] by permission of the author and the American Chemical Society; copyright ⓒ 1973.)

the practical detection limit is limited less by the sensitivity of the spectrometer than by difficulties involved in sample handling. However when high sensitivity is needed the use of a Fourier transform spectrometer will give somewhat better results than a good grating spectrometer.

IV.   THE SPECTROSCOPY OF TRANSIENT SPECIES

Owing to the very short measurement time made possible by the use of rapid-scanning interferometers, absorption spectra of species that are only present in the infrared beam for a few seconds can be easily measured. By speeding up the mirror velocity, as many as 100 scans/sec have been measured using commercially available instrumentation. The most important application of the rapid-scanning capability for analytical chemists is undoubtedly the measurement of the absorption spectra of fractions eluting from a gas chromatograph without trapping the sample. The next section is devoted entirely to a description of this technique.

One of first field measurements made using an interferometer demonstrates the power of this technique for measuring the spectra of transient species. Spectra of a high altitude nuclear fireball were measured [18] from an aircraft using both an early Block Engineering rapid-scanning interferometer with a 1 cm aperture and an Ebert-Fastie spectrometer with a 20×30 cm grating. The grating spectrometer measured no data at all, while features due to CO, NO, and NO^+ could be seen in the interferometrically measured spectra (Fig. 10.12). By comparison with synthetic spectra, the vibrational and rotational temperatures of these molecules could be estimated. The structure of the first CO overtone was matched with a theoretical spectrum where the vibrational and rotational temperature were both equal to 3500 K.

One of the few published examples of absorption spectroscopy of reacting species was described by Low, Epstein, and Bond [19] who showed that previously published spectra of some methyldiboranes were incorrect; the species that had been initially prepared had reacted by the time it took to measure the spectrum on a grating spectrometer. Low and his co-workers carefully fractionated methylated diborane equilibrium mixtures to yield fractions on the order of 10^{-7} mole. A selected well-resolved fraction was frozen out in a liquid-nitrogen colled capillary attached to a small infrared cell. Spectra were measured at 18 cm^{-1} resolution using a Block interferometer; by measuring the spectra at one minute intervals it could be shown how the absorption bands varied in intensity and frequency with time. For instance, a band near 1600 cm^{-1} previously ascribed to a B–H (bridged) stretching mode decreased significantly in intensity which was taken to indicate a disruption in the diborane skeleton. The way in which the spectrum of one of these compounds varies is shown in Fig. 10.13. The first spectrum to be measured, A, is thought to be a good approximation of the true spectrum while the final two spectra, F and G, are very similar to the previously published spectra of this compound.

In a later publication, Low, Epstein, and Bond [20] measured the spectra of a series of methylated diboranes by the same method. They showed that although there was good correspondence between the rapidly measured interferometric spectra and some earlier grating spectra, the correspondence decreases with the increasing extent of methyl substitution. From this work, these authors were able to distinguish some characteristic vibration frequencies of the methyl diboranes.

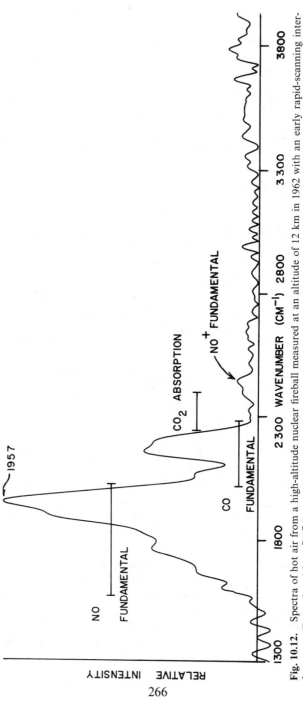

Fig. 10.12. Spectra of hot air from a high-altitude nuclear fireball measured at an altitude of 12 km in 1962 with an early rapid-scanning interferometer. (Reproduced from [18] by permission of the author).

266

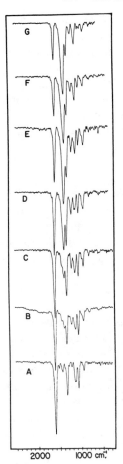

Fig. 10.13. The effect of time on the absorption spectrum of cis-1,2-dimethyldiborane. The total time that the sample was in the gas cell was: *A*, 1 min; *B*, 4 min; *C*, 7 min; *D*, 10 min; *E*, 13 min; *F*, 20 min; *G*, 40 min. (Reproduced from [19] by permission of the author and the Chemical Society; copyright © 1967.)

The compounds studied by Low and his co-workers decomposed slowly compared with the time taken to scan the complete interferogram. For reactions in which the half-life of the reactants is of the same order as the scan time of a typical rapid-scanning interferometer, the spectral features of the reactant molecules may show some broadening. For example, consider a reacting molecule in the spectrum of which there appears a very sharp spectral line. If no reaction occurred, the interferogram due to this line would be sinusoidal, and the width of the line after the Fourier transform would be approximately equal to the width of the ILS. On the other hand, if the concentration of the reactant decreased exponentially during the measurement, the computed line would have a Lorentzian line shape whose width would be *greater* than the width of the ILS.

Thus to obtain accurate spectral data during a chemical reaction one must ensure that the scan time of the interferometer is short compared to the reaction half-life. To achieve this goal, it may become necessary to increase the scan speed of the interferometer. Since increasing the scan-speed has the effect of changing the audio-frequency limits of the interferogram, one must ensure that not only is the audio-frequency pass-band of the amplifier changed but also that the response time of the detector is sufficiently short that it can follow the high frequency modulations in the interferogram.

V. GC–IR

Undoubtedly the greatest single chemical application for rapid-scanning interferometers in the mid-infrared involves the on-line identification of fractions separated by a gas-chromatograph (GC–IR). The importance of this technique was recognized only a short time after the initial development of rapid-scanning interferometers, but it has not been until the development of sophisticated data-systems to handle the tremendous amount of data that is produced during GC–IR that operation of such instruments can be performed in a routine fashion.

The first study investigating the feasibility of using a rapid-scanning interferometer to measure the spectrum of gas-chromatographic peaks was carried out by Low and Freeman [21] in 1967 using a low resolution Block Engineering interferometer together with an analog tape-recorder. A sequence of interferograms was recorded during the entire chromatogram. At the end of the run a single interferogram or any number of consecutive interferograms were fed to a computer for addition, storage and subsequent data processing. Spectra were measured in a "pseudo-transmittance" format by subtracting from the single-beam spectrum of the cell with the sample present an equal number of co-added scans of the spectrum of the empty cell, and presenting the spectrum inverted. In this way the spectrum is presented in the same format as a transmittance spectrum so that it was easier for chemical spectroscopists to interpret. Spectra were measured with the effluent peaks trapped in or flowing through the gas-cell which was 5 cm long and 4 mm in diameter.

Using a synthetic mixture containing approximately equal portions of six volatile liquids, Low and Freeman were able to demonstrate the feasibility of measuring spectral bands from methyl acetate samples as small as 5 μg provided that the sample was trapped in the gas-cell and extensive signal-averaging (300 scans) was performed. One interesting experiment which was performed during this study was to measure the interferograms corresponding to the leading edge and the trailing edge of a single unresolved GC peak containing iso-octane and methyl ethyl ketone (Fig. 10.14). It is easily seen that the leading edge (spectrum A) consists primarily of iso-octane while the

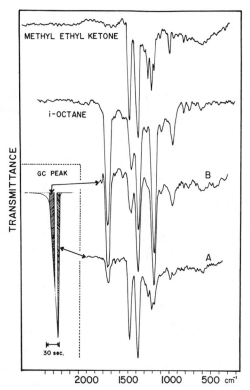

Fig. 10.14. Spectra of *trapped* fractions from an unresolved GC peak containing both methyl ethyl ketone and iso-octane. (Reproduced from [21] by permission of the author and the American Chemical Society; copyright © 1967.)

trailing edge is mainly methyl ethyl ketone, although evidence of the minor component is seen in each spectrum.

The sensitivity of this arrangement for flowing samples was naturally considerably below that for trapped samples, since only a few scans could be taken. From the published spectra of Low and Freeman the detection limit for samples measured in this way was generally a little under 100 μg. Part of the reason why fairly noisy spectra were measured in this experiment was the low temperature of the infrared source (700°C); presumably the temperature was kept low to avoid the effects of digitization noise. In a later study, Low [22] investigated the use of dual-beam techniques for the measurement of gaseous fractions separated by a gas-chromatograph. His experimental arrangement was shown earlier in Fig. 7.2. The light-pipe gas-cells used by

Low in this study had a volume of 22 ml, which is rather large for GC–IR, but Low was still able to demonstrate an increase in sensitivity over the previous method for the measurement of trapped samples. Griffiths and Lephardt [23] used smaller gas-cells (approximately 1 ml in volume and 5 cm path-length) and a different method of producing the optical null (see Fig. 7.3), and measured spectra at approximately the same sensitivity as Low. In retrospect, it seems that the light-pipe used by Griffiths and Lephardt was too small for efficient use for GC–IR under the conditions of their experiment while the cell used by Low was probably too big. The dimensions of the cell used in GC–IR are very critical to the success of the technique and are strongly dependent on whether the sample is to be studied while it is trapped in the light-pipe or if it is flowing through it. The optimum dimensions for a GC–IR light-pipe will be discussed later in this section.

The development of high resolution interferometers with computerized data systems has led to great advances in GC–IR. While the advances have been in part due to an increase in sensitivity through more advanced optical technology, more important is the increased data-handling capability. Spectra measured at 4–8 cm^{-1} resolution are adequate for most GC–IR applications, which is the resolution at which many spectra in the laboratory are measured on grating spectrometers. Figure 10.15 shows two superimposed ratio-recorded polystyrene spectra measured at this resolution under the same type of experimental conditions that would be encountered in GC–IR, that is, only a short time available for measurement of the sample spectrum but a long time available for the measurement of the reference. The spectra of polystyrene were each measured using 1 scan and the single-beam

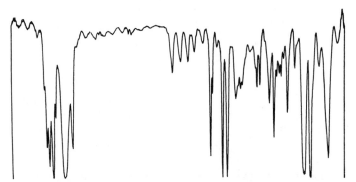

Fig. 10.15. Two superimposed spectra of polystyrene, each measured using 1 scan through the sample and ratioed against the same (16 scan) reference. This spectrum demonstrates the repeatability and low noise level now able to be achieved with modern Fourier transform spectrometers in very short measurement times.

spectra run in this way were ratioed against a reference spectrum measured using 16 scans. In practice, noise levels this good are not able to be achieved for GC–IR using one scan because the light-pipe and its condensing optics generally do not transmit a high proportion of the incident radiation. One real advantage of a ratio-recording system for GC–IR is the very flat baseline which can be obtained by using an identical optical path for the sample and reference spectra, so that large ordinate scale expansion over extended frequency ranges becomes possible.

The first work using a computerized Fourier transform spectrometer (the Digilab FTS–14) was performed by Lephardt and Griffiths [24] using a light-pipe of 3 ml volume and 6 mm diameter. The sensitivity of this spectrometer was such that several bands of acetophenone (a fairly strong infrared absorber) when only 2 μg were in the light-pipe as the sample was flowing through. However, the width of the GC peak for this experiment was such that it required a total injection of about 10 μg of acetophenone so that at any instant 2 μg was present in the light-pipe. Thus the sensitivity of this measurement was only slightly better than those of the low-resolution measurements described earlier. This study showed the importance of matching the volume of the light-pipe to that of the GC peak.

The optimum dimensions of a light-pipe are strongly dependent on whether the sample is to be trapped in the light-pipe for the length of time required for the measurement of the spectrum or whether the carrier gas is constantly flowing through the cell. If trapping techniques are employed then it is important that all the sample in a given GC peak is present in the light-pipe. In this case, a system such as that manufactured by the Wilks Scientific Corporation [25] (Model 41B Vapor Phase GC–IR Analyzer) is desirable. The light-pipe of this device is of small volume (0.6 ml) and relatively long pathlength (6 cm) so that a given amount of material in this cell produces stronger absorption bands than the same amount of material trapped in a cell of larger volume or shorter pathlength. However the volume of this light-pipe is smaller than the volume of carrier gas containing most gas-chromatographic peaks. To ensure that *all* the sample is trapped inside the light-pipe, the carrier gas is passed through a short length of tubing which is coated with a typical liquid phase used in gas chromatography. When the GC peak of interest is detected this length of tubing is rapidly cooled so that all of the sample is condensed on the walls of the tubing. After all the sample contributing to one peak has been trapped, the tubing is rapidly heated and the sample is flushed into the light-pipe and trapped there. In this way a GC peak which has a wide half-width, and hence elutes in a considerably larger volume of carrier gas than the volume of the light-pipe, can be concentrated and trapped in the light-pipe.

Low and his co-workers [12, 26] have used this technique to measure the spectra of several organic compounds by Fourier transform spectroscopy. He showed that strong spectra of microgram amounts of some materials could be measured and that some bands of submicrogram quantities could be detected using this technique (Fig. 10.16). He used a Wilks Model 41B with its optics modified to be compatible with the circular beam of the interferometer. His sampling system is shown in Fig. 10.17. The mirror T was a 90° toroid which was rotated 90° through its optical axis. Rather than giving a small circular focus, the use of the toroid in this fashion distorted the beam to produce a focus of the same rectangular dimensions as those of the light-pipe, thereby increasing the percentage transmission of the cell.

When flow-through measurements are to be made, the dimensions of the light-pipe must be different from the light-pipe described above. In this case, matching the volume of the cell to the volume of the GC peak is the most important consideration, since sample concentration techniques cannot be applied. The volume of the light-pipe, V, should be approximately equal to the average half-width of the peaks being studied. The optimum sensitivity of a cell of volume V is achieved when the cross-sectional areas, A, is made as small as possible and the pathlength, b, is increased accordingly, while keeping the transmission, T, of the cell as large as possible. Obviously A cannot be decreased indefinitely or the transmission of the cell will be reduced drastically and insufficient energy will reach the detector. There is, therefore, a tradeoff between three factors, A, b, and T.

In practice the cross-sectional area of the cell cannot be reduced much below 0.1 cm^2 since the beam in the sample compartment of a spectrometer cannot be condensed by more than a 4x beam-condenser without losing too much light by reflections from the walls of the light-pipe. The maximum sensitivity for GC–IR is achieved by ensuring that the halfwidth of the GC peak is small, so that for a light-pipe of a given cross-sectional area the length of the light-pipe (and hence the number of reflections occurring as the beam is transmitted through the light-pipe) is as small as possible. To achieve a small peak half-width using conventional packed columns on the chromatograph can present a problem even for the most experienced chromatographers. Typical parameters when packed columns are used are 20 sec half-width with a carrier gas flow-rate of 30 ml/min. This means that the volume of carrier gas containing the GC peak between its half-width points is 10 ml. If a light-pipe of 0.1 cm^2 area is used for the measurement of this peak, it would need to be 1 m in length to accept the complete sample. It is unlikely that the light-losses in such a cell would permit enough energy to reach the detector that spectra could be measured rapidly without the use of a cooled detector. Methods of reducing the volume of each peak therefore have to be employed.

One method is to increase the temperature of the GC column; this tech-

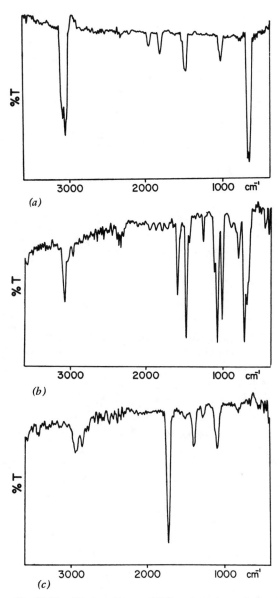

Fig. 10.16. Spectra of *trapped* GC peaks measured after concentrating the sample; each spectrum was measured with a 0.001 μl sample trapped at 150°C in a Wilks light-pipe, using 120 scans. Upper trace, benzene; middle trace, monochlorobenzene; lower trace, 2,4,6-trimethylpyridine. (Reproduced from [12] by permission of the author and the Institute of Petroleum; copyright © 1972.)

273

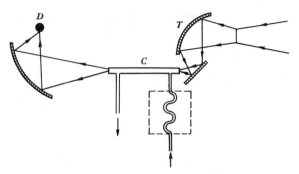

Fig. 10.17. Experimental arrangement for concentrating the GC effluent and then passing it into the light-pipe. *T* represents the toroidal mirror used to distort the beam to the configuration of the light-pipe. (Reproduced from [26] by permission of the author and the Federation of Societies for Paint Technology; copyright © 1971.)

nique usually results in a loss of resolution and has one other disadvantage in that most liquid phases cannot be taken above a certain temperature without decomposition or evaporation. A better method may be to use a support-coated open tubular (SCOT) [27] column for the chromatography. Although the capacity of SCOT columns is not as high as that of conventional packed columns, the resolution is much higher and the volume of carrier gas containing the peak is also reduced. In this way the peaks may be sharpened up so that their half-width is typically one or two milliliters.

Let us compare the sensitivity able to be achieved for a GC peak contained in 5 ml of carrier gas using three commercially available light-pipes of different dimensions, assuming the same transmittance for each light-pipe. The dimensions of the light-pipes are shown in Table 10.1.

TABLE 10.1

Manufacturer	Length (cm)	Area (cm^2)	Volume (cm^3)
Wilks	6	0.09 (1.5 × 6 mm)	0.54
Digilab	5	0.28 (6 mm diameter)	1.4
Norcon [28]	30	0.16 (4 × 4 mm)	4.8

Since the absorbance, $A(\bar{v})$, of any band with an absorptivity $a(\bar{v})$ is given by

$$A(\bar{v}) = -\log_{10} T = a(\bar{v}).b.c$$

and since $a(\bar{v})$ and c (the concentration of the sample in the carrier gas) are constant, the absorbance of the sample contained in a large volume of carrier gas is going to be five or six times higher using the longer Norcon light-pipe than if either of the other two light-pipes were used. While the transmittance through the longer light-pipe is probably going to be smaller than that through the larger area Digilab cell, it is unlikely to be down by more than a factor of two since the Norcon cell is used in their rapid-scanning dispersive spectrometer where energy is almost certainly at a premium. The Wilks light-pipe is best used under conditions where c has been made as large as possible and so should not be considered for flow-through measurements unless exceptionally narrow peaks are being studied.

This example demonstrates the real importance of light-pipe design in GC–IR. Let us now estimate the practical sensitivity of a GC–IR system where the light-pipe has the same dimensions as the Norcon device. The sensitivity is strongly dependent on the type of sample being measured. For his collection of vapor-phase spectra, Welti [29] has employed a gas-cell of 25 ml volume and 9.2 cm path-length. The amount of sample required to measure a full strength spectrum varied considerably. For example, for the measurement of the spectrum of salicylaldehyde only 0.5 μl of liquid were needed to obtain a spectrum where ten bands in the "fingerprint" region below 2000 cm^{-1} absorbed more than 50% of the radiation (absorbance greater than 0.3 absorbance units). The carbonyl band at 1680 cm^{-1} shows an absorbance of approximately 2 (Fig. 10.18a). On the other hand, 6 μl of 1-decene were required in order that two bands absorbed more than 50% in the same region (Fig. 10.18b). Thus the sensitivity of any GC–IR system would need to be at least ten times greater for 1-decene than for salicylaldehyde.

If bands having an absorbance of 0.02 (5% absorption) can be easily distinguished from the noise in the GC–IR experiment, it can be seen that 10 bands of salicylaldehyde could be measured if

$$0.5 \times \frac{4.8}{25} \times \frac{9.2}{30} \times \frac{0.02}{0.3} \; \mu l$$

or 0.002 μl (approximately 2 μg) were present in the light-pipe. If the carbonyl band is the only feature of interest it could be detected if less than 300 nanograms is present in the light-pipe. On the other hand, the detection limit for the few bands of 1-decene below 2000 cm^{-1} is greater than 10 μg.

In summary, it can be seen that, since bands absorbing considerably less than 5% can be observed in practice, the detection limit of a well-designed GC–IR system should be in the submicrogram range if polar molecules are being studied, but somewhat larger for non-polar molecules.

Although several workers described early experiments where spectra of GC peaks were measured without trapping the sample, many of these investigations have only been performed with difficulty due to the problems involved in handling the data and controlling the experiment, and merely

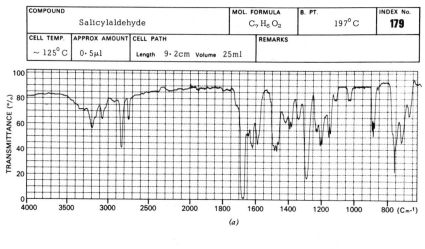

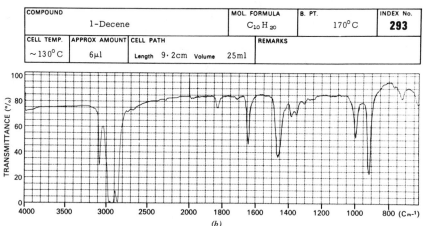

Fig. 10.18. Infrared reference spectra of salicylaldehyde and 1-decene vapors measured in the same gas cell. Note that over 10 times as much 1-decene was needed to obtain a spectrum with weaker absorption bands below 2000 cm⁻¹. (Reproduced from [29] by permission of Heyden & Son Ltd.; copyright © 1970.)

demonstrated the feasibility of the method. Digilab developed the first completely automated GC–IR system designed for flow-through measurements. Their system is an integrated package of optics, electronics, and software which nicely illustrates both the power of Fourier transform spectroscopy for measuring infrared spectra rapidly and the power of a computer for rapid data collection and hardware control. The system is basically in accessory to the FTS series spectrometers: the optics fit directly into the sample compartment of the instrument, the electronics are contained in a rack that is mounted under the controller electronics of the interferometer and the software is stored directly onto the disc where the control programs for the spectrometer are stored.

The central component of the optics is the light-pipe (see Table 10.1) through which the effluent gas from the chromatograph is passed. The light-pipe is held in an oven so that its temperature can be set an any value between ambient and 400°C. The infrared beam is passed through the light-pipe by means of a 2x beam-condenser, so that when the system is well aligned a little less than 50% of the light from the source reaches the detector. A heated valve allows the gases inside the light-pipe to be trapped while the gas still eluting from the chromatograph bypasses the light-pipe. In this way extended signal averaging can be used to increase the signal-to-noise ratio of the spectrum. If several GC peaks are to be identified, each of which contains so little sample that trapping could optimally be used for the spectral measurements, then the interrupted elution technique described by Scott et al. [30] can be employed. Here not only is the gas in the light-pipe trapped but a valve is also operated to shut off the flow of carrier gas through the column. It has been shown that, provided that the carrier gas is not kept stationary for excessively long periods of time, little degradation in the concentration profile of each peak will be noted. After the desired amount of signal averaging has been carried out, the carrier gas flow is automatically restarted both through the GC column and through the light-pipe.

The gases are passed into the light-pipe from the chromatograph by means of a heated copper-clad stainless-steel tube (1/8 in. o.d.). Nondestructive GC detectors such as the thermal conductivity type are preferable for this application since the sample can be passed directly from the detector to the light-pipe. If a destructive detector, such as the flame ionization type, is used a splitter must be employed so that at least 50% of the sample by-passes the detector and travels directly to the light-pipe from the column.

The signal from the GC detector is monitored to detect when it reaches a certain threshold voltage or falls below it. When the signal goes above the threshold voltage, there is a delay of t sec to allow the sample to travel from the detector to the light-pipe after which interferograms are signal-averaged until t sec after the GC signal falls below the threshold level. The signal-averaged interferogram is then moved to a certain address in the disc memory and the spectrometer waits until the next peak in the chromatogram. At this point the same process is repeated with the interferogram being stored at the address in the computer memory adjacent to that of the previous interferogram.

The delay time, t, which allows the peak to travel between the detector and the light-pipe may be determined manually or automatically. In both methods a small volume of a pure solvent is injected into the chromatograph and the time taken for the infrared absorption bands due to the solvent to reach their full intensity is noted. When the manual method is used, the delay time is varied for a series of injections and the absorbance of a band in the solvent spectrum is plotted out. Low [31] has demonstrated this method using a wide GC peak (Fig. 10.19). It is seen from this series of spectra that a 20-second delay time was needed for the chromatographic conditions used for this experiment. Since t is only dependent on the volume between the GC detector and the light-pipe and the carrier gas flow-rate, t remains constant for all chromatographic runs performed with the same flow-rate.

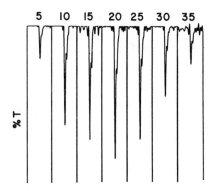

Fig. 10.19. Calibration procedure to determine the delay time, t. Aliquots are injected into the GC and at known times after the maximum signal at the GC detector, 1 scan is taken. The spectrum is calculated and a small region in which a strong band is known to absorb is plotted out. The procedure is repeated for a range of times and the time for which the maximum absorbance is observed gives t. (Reproduced from [31] by permission of the author and the American Chemical Society; copyright © 1971.)

The automatic method is quicker but is sometimes less accurate. This technique uses some electronics which have been devised to measure the rate of change of the GC signal and show when the signal is at a maximum. At this point the interferometer is set scanning at one second intervals, and the amplitude of the interferogram at zero retardation is monitored. When the sample is present in the light-pipe at its maximum concentration, the amplitude of the interferogram at zero retardation will be a minimum due to absorption by the sample. The number of scans taken to reach this condition is recorded by the computer and used as the value for t.

Ratio-recorded spectra are measured by recording a background spectrum through the empty light-pipe before or after the sample chromatogram. This background spectrum is measured using a larger number of scans than were used for the widest GC peak (at least four times as many) so that the noise-limiting factor is the short time that each GC peak is present in the light-pipe. The interferogram of each GC peak is then successively recalled, the spectrum is computed and ratioed against this background. In this way very flat baselines are obtained which are suitable for large ordinate scale expansion.

Using this system, spectral bands from as little as 1 μg of material have been able to be measured without trapping the sample [32]. If 20 to 40 μg of sample is present in the GC peak, most of the characteristic bands can be easily seen in the spectra. Spectral differences from such chemically similar molecules as methyl caprate and methyl caprylate are able to be seen in spectra measured using this system [33] (Fig. 10.20). The differences between these spectra are primarily seen in the relative intensities of the absorption bands since the principal difference between these molecules is the length of the alkyl group. When spectra of GC peaks are measured interferometrically, it is usual that several interferograms are averaged over a single GC peak; in this way the resultant intensities of the bands are the average of the intensities for each scan. For the low concentration samples usually encountered in GC–IR, spectra measured in this way show little distortion from the true spectrum. If a rapid-scanning monochromator were used for the measurement using a single scan across the entire GC peak, the bands at the beginning and end of the spectrum would show a low absorbance relative to the bands at the center of the spectrum where the concentration of the sample in the light-pipe was highest.

Recently a dual-beam rapid-scanning spectrometer was introduced by Spectrotherm Corporation [24], the primary purpose of which is the on-line identification of GC peaks. The optical layout of this spectrometer was described in Chapter 7, Section I, and the data system was briefly discussed in Chapter 6, Section IV. The high sensitivity of this instrument is derived from the use of a very sensitive MCT detector, the D^* of which is almost 100 times greater than that of the TGS detectors commonly used with other rapid-scanning interferometers. If such a sensitive detector were used with an interferometer operating in the conventional single input beam/single output beam mode of FTS, the S/N of the interferogram would exceed the dynamic range of the ADC, so that it is easy to see why the dual-beam optics are so important for the success of this instrument.

Another innovation seen in this system is the design of the gas-cell, the optical diagram of which was shown in Fig. 7.5. It can be seen that the only reflections taking place in this cell are from the mirrors at either end of the cell, and no reflections at the wall are allowed. This arrangement has certain advantages in terms of energy throughput, but leads to rather poor flow characteristics, so that when the carrier gas continually flows through the cell, traces of a sample may be detected spectroscopically a few seconds after all the material should have passed through the cell. To alleviate this situation, an ingenious flushing device has been devised to permit the cell to be completely flushed between *resolved* GC peaks. For unresolved peaks, a spectrum of the leading edge of the first component may be measured with no signal averaging—this is possible because the operator is able to view individual

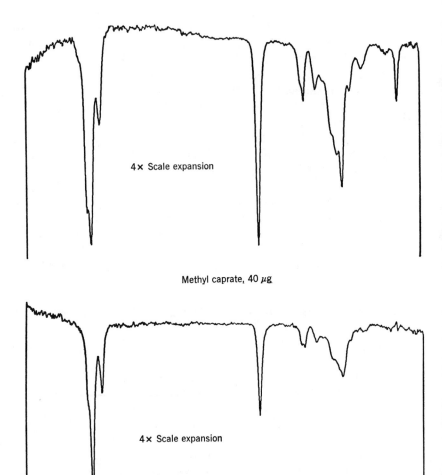

4× Scale expansion

Methyl caprate, 40 μg

4× Scale expansion

Methyl caprylate, 40 μg

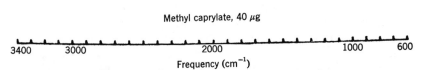

3400   3000   2000   1000   600

Frequency (cm⁻¹)

Fig. 10.20. Spectra of 40 μg of methyl caprate (upper trace) and methyl caprylate recorded *without trapping* using a Digilab GC–IR system. A good identification of the samples can be made using these spectra in conjunction with retention time data. (Reproduced from [33] by permission of Heyden & Son Ltd.; copyright © 1972.)

spectra on an oscilloscope display just two seconds after the end of the corresponding interferogram. This spectrum may be stored and later used to subtract out bands due to the first component from a signal-averaged spectrum in which bands due to both components of the chromatographic doublet are seen. Although few results are presently available, the Spectrotherm spectrometer may well represent the first GC–IR system whose sensitivity is sufficiently low that GC–IR can be a useful technique for organic and analytical chemist dealing with submicrogram quantities of materials.

Another method of measuring the infrared spectra of GC peaks has been suggested by Lephardt and Bulkin [35]. The heart of their system is a cholesteric liquid crystal film. The interface between the film and the GC effluent operates on the same principle as the gas chromatograph that precedes it. A thin film of a liquid phase material is coated on an infrared transmitting window so that as the effluent from the GC detector flows over it, some of the sample dissolves in the liquid. If the film transmits enough infrared energy, a spectrum of each component in solution in the film may be recorded as the zone of the dissolved sample moves across the film.

The choice of film material is crucial to the success of this technique. Lephardt and Bulkin have listed the five criteria that they consider a liquid should meet for this application. Three of these are fairly obvious: the liquid should be chemically inert, stable over a wide frequency range, and show a high transmission over the complete frequency range. However, the final two, that the film must efficiently trap and concentrate a wide variety of organic compounds in a nonselective fashion and that the film must be able to regenerate quickly in the flow of carrier gas, appear to be contradictory. If the film is able to regenerate quickly there seems little likelihood that the sample will be concentrated by the liquid phase. If the sample does dissolve for a sufficient length of time for it to become more concentrated, then there is little chance that it will be able to be rapidly regenerated by the carrier gas. Thus either *very* finely controlled temperature programming would have to be performed for this cell or the two criteria would not be able to be met simultaneously.

Lephardt and Bulkin have listed several advantages for their technique. The spectra that are measured are solution phase spectra and may therefore be directly compared with standards. By using cholesteric liquid crystals, many samples cause color changes on solution and these color changes can be used to trigger data collection on the interferometer. Thus an accurate determination of the time taken for the sample to reach the cell from the GC detector is not necessary. However, the type of liquid crystal used by Lephardt and Bulkin had good spectral windows only in regions of low spectroscopic interest; in several important regions (particularly the ones around 1700 and

2900 cm^{-1}) the transmittance was less than 10%. It is difficult to imagine any liquid having good spectral windows in all regions of importance for GC–IR. The spectra shown by these authors demonstrate many of the advantages and pitfalls of their technique. The profiles of the absorption bands are undoubtedly more similar to liquid phase spectra that to gas-phase spectra. However, the sensitivity is not as high as published results using gas-phase sampling imply, while evidence of the sample adhering to the liquid phase for long periods of time has also been shown (Fig. 10.21). Thus it appears that gas-phase techniques are the most promising for GC–IR. However, the method of Lephardt and Bulkin might have potential importance for detecting and identifying high molecular weight compounds present in the atmosphere at very low concentation.

VI. TRACE GAS ANALYSES

Most of the earliest attempts to detect trace components in gaseous samples by interferometry involved the use of dual-beam Fourier transform spectro-

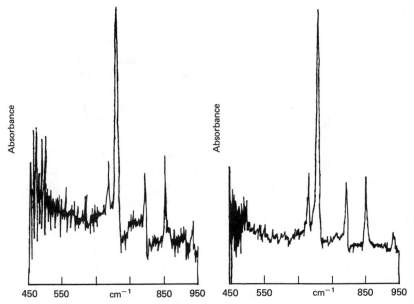

Fig. 10.21. Spectra of nitrobenzene recorded using the liquid crystal GC–IR cell, (a) immediately upon entrance of the sample and (b) approximately 30 min later. In order to remove the sample completely from the liquid crystal, Lephardt and Bulkin had to heat the cell above the cholesteric–isotropic transition temperature and flush the liquid crystal with air for about 2 min. (Reproduced from [35] by permission of the author and the American Chemical Society; copyright © 1973.)

scopy so that digitization noise was not the limiting factor for the sensitivity. Bar-Lev [36] demonstrated that by using an early Block Engineering interferometer in conjunction with a 10 m path-length gas-cell, as little as 10 ppm of methanol is detectable with only one scan. Using this arrangement it should have been possible to detect 1 ppm of methanol by multiple scanning and a smaller quantity of strongly absorbing compounds like the Freons. Low and Mark [37] also used a dual-beam arrangement to detect SO_2 in a small gas-cell (0.25 cm^3 volume, 1 cm path-length). Strong spectra of SO_2 were measured when the pressure in the cell was 0.4 torr (Fig. 10.22) and, using these data and some reasonable assumptions, Low and Mark estimated that the sensitivity of dual-beam Fourier transform spectroscopy for SO_2 detection could be pushed into the parts-per-billion range.

With the advent of higher resolution interferometers and computerized data systems, even higher sensitivity is now able to be achieved without the need of resorting to dual-beam techniques with the concomitant experimental difficulties. Much of the work on trace gas analysis using FTS has been carried out by Hanst's group at the U.S. Environmental Protection Agency's Air Pollution Research Center. Hanst, Lefohn, and Gay [38] have discussed how the sensitivity of infrared absorption spectroscopy may be increased; the detection sensitivity of an experiment depends on:

(a) the strength of the absorption bands;
(b) the noise in the spectral background;
(c) the structural detail in the pollutant bands;
(d) the strength and structure of the bands of interfering gases.

Since air pollutants are generally present in low concentration, their infrared absorption is extremely weak over short paths so that long path cells have to be used in air pollution spectroscopy. Hanst and his co-workers have designed a cell in which a path-length up to 720 m can be obtained. The multiple-pass cell used initially was a three mirror cell of the type described by White [39], although later experiments by this group have used an eight mirror cell that offers greater optical throughput. For either cell the light losses occurring on reflection and transfer are quite large so that the use of cooled detectors (in particular the liquid-nitrogen cooled mercury cadmium telluride photodetector) is beneficial, and in Hanst's system the digitization noise limit was not reached. The capability of measuring spectra with a low background noise level is necessary in order that high resolution spectra may be measured so that the effect of interfering gases is reduced.

Lines due to interfering small molecules which overlap with the lines in the sample spectrum at low resolution will become separated as the resolution is increased. Naturally if there is no rotational fine structure there is nothing to be gained by increasing the resolution. Often the oxygen and nitrogen

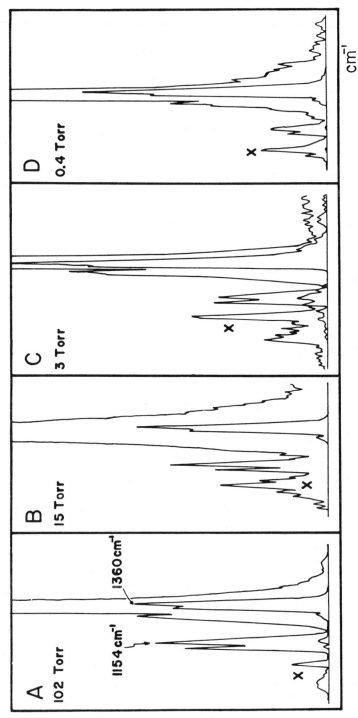

Fig. 10.22. Spectra of SO_2 measured by dual-beam FTS using a 1-cm path-length cell. The SO_2 pressure in torr was: *A*, 102; *B*, 15; *C*, 3; *D*, 0.4. *X* represents spectral structure resulting from uncompensated instrument background. (Reproduced from [37] by permission of the author and the D. Reidel Publishing Co.; copyright © 1971.)

284

present in the atmosphere cause pressure broadening of the lines, and reduction of the total pressure of the gas will allow fine structure to be resolved in the spectrum. There is obviously a trade-off between selectivity and sensitivity, since reduction of the pressure removes some pollutant from the light path. The optimum pressure at which to perform this measurement is determined by the complexity of the bands in the spectral region under study.

The principal interference in the infrared spectrum of atmospheric pollutants is water vapor, and its effects must be reduced as much as possible. Measuring linear transmittance spectra by ratio-recording rather than an optical null has a great advantage in that only one cell is needed for the measurement, and the spectrum of an uncontaminated air sample may be measured before or after the measurement of the contaminated sample. The availability of mass memory data systems such as discs or tapes means that the spectrum may be measured at different partial pressures of water vapor until an exact match is found. It should also be possible to measure just one reference spectrum of water vapor and adjust the absorbance of the bands digitally.

Hanst et al. have demonstrated the advantages of removing the bands of water vapor from the spectrum. Figure 10.23 shows the single-beam spectra of contaminated and pure air and the result of ratioing these two spectra. Obviously certain of the water lines are so strong that the energy transmitted at their peak is so small that at that frequency almost full-scale noise is seen. It is preferable to lift the recorder pen when the energy in the reference spectrum drops below a certain level and move quickly to the next frequency at which there is sufficient energy as described in Chapter 6, Section IV, so that "cleaner" transmittance spectra than the one shown in Fig. 10.23 can be measured. Even so, these spectra clearly demonstrate the power of high resolution Fourier transform spectroscopy for measuring weak bands from pollutants whose strongest bands occur where water vapor has strong absorption features.

Even in spectral regions that are commonly considered as spectral "windows" the effects of absorption lines due to carbon dioxide and water vapor can often be seen when very long absorbing paths are studied. Hanst has shown a single-beam spectrum of a sample of air collected from close to a highway. The principal lines in this spectrum (Fig. 10.24) are due to H_2O and CO_2 but by ratioing this spectrum against one of uncontaminated air with the partial pressure of H_2O and CO_2 adjusted to give good cancellation of their rotational fine structure, features due to ethylene (0.19 ppm), acetylene (0.25 ppm), propylene (0.12 ppm) and formic acid (0.16 ppm) are easily seen. The path-length for this measurement was 418 m and the sample was "diluted" with an equal volume of pure air.

The actual resolution required to give the optimum sensitivity depends on

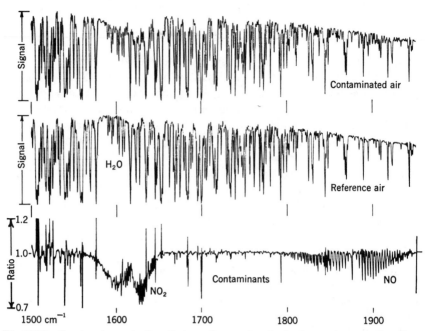

Fig. 10.23. Spectra demonstrating that results can be obtained even in spectral regions where strong atmospheric absorption occurs. The air pressure in the cell was 150 torr and the path length was 81 m; the NO_2 concentration was about 5 ppm, and the NO concentration was about 10 ppm. (Reproduced from [38] by permission of the author and the Society for Applied Spectroscopy; copyright ⓒ 1973.)

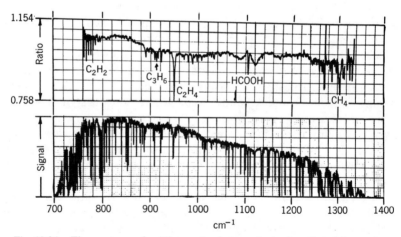

Fig. 10.24. The spectrum of a 418-m path of atmospheric air collected from near a highway, plotted in the region from 700 to 1400 cm^{-1}. The lower trace is the single-beam spectrum of the sample and the upper trace is the result of ratioing this spectrum against a reference containing the same amount of water vapor. (Reproduced from [38] by permission of the author and the Society for Applied Spectroscopy; copyright ⓒ 1973.)

286

the amount of fine structure in the spectrum of the species of interest. Thus Hanst found that in measuring the spectra of 30 ppb of ozone and nitric acid and 3 ppb of ethylene, 8 cm^{-1} resolution was required for ozone, 2 cm^{-1} resolution was needed for nitric acid while about 5 cm^{-1} resolution gave the optimum results for ethylene. The spectrum of ethylene is shown in Fig. 10.25 to demonstrate the magnitude of the ordinate scale expansion which can be used under the low energy conditions of this experiment. The absorption

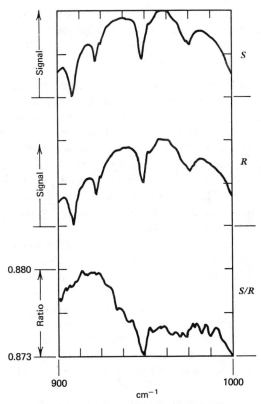

Fig. 10.25.   The spectrum of 3 ppb of ethylene in room air; the path-length is 471 m and the resolution was 2 cm^{-1} (with smoothed plotting). S is the scale expanded single-beam spectrum of the sample and R is the corresponding spectrum of the reference air, while S/R is the ratio of these two spectra. Note that the ordinate scale expansion is greater than 100x. (Reproduced from [38] by permission of the author and the Society for Applied Spectroscopy; copyright © 1973.)

band under study is only absorbing approximately 0.25% of the incident radiation.

Using the approaches described previously by Stephens [40] and Hirschfeld [41], Hanst et al. [38] have been able to estimate the detection limit for many common atmospheric pollutants, C_{min}, as:

$$C_{min} = \frac{(1 - P) \cdot N_1}{0.37 \cdot a \cdot L}$$

where P is the cell transmittance per pass;
    N_1 is the noise level, as a fraction of the total signal, after one pass;
    a is the absorption coefficient, the values of which are usually in the range 10–50 per cm per atm;
    L is the pathlength of the cell per pass.

N_1 is kept to a minimum by using a Fourier transform spectrometer and a cooled detector. Hanst's group used a rather large cell where $L \sim 10$ m, but they were able to forecast that for monitoring urban air an absorption cell much smaller than 10 m long could be used. Table 10.2 shows the detection limits that are calculated for a cell in which 50 passes through a 10 m pathlength cell are made, assuming a noise level the same as that of Hanst's system, with $P = 0.98$. The limit is below 1 ppb in most cases. Even if one allows a factor of 10 degradation of sensitivity due to interferences and other practical effects, the detection limit of most compounds is still below the level of ambient atmospheric pollution.

This system has been found to be so promising for studying atmospheric pollution that it is being used in laboratory research and for field measurements. In the laboratory photochemical reactions are being studied for such systems as NO_2–ethylene–air and auto-exhaust–air. Measurements can be carried out sufficiently quickly that the increase and decrease in concentration of important intermediates in photochemical reactions can be studied. A similar system with a shorter path-length gas-cell is being installed in a van for mobile measurements of ambient air while a further system is being used to provide a sensitive analysis of Southern California photochemical pollution [42].

The interferometer described by Schindler [43] and discussed in Chapter 4, Section 1.D has been used for several studies of gases over long absorbing paths. For example, by placing an infrared source on one side of a highway and the interferometer on the other side, Farmer [44] was able to study the variation of the concentration of pollutant gases in the atmosphere over the course of several days. The same instrument has been used for measuring the composition of the stratosphere using the sun as source [45, 46]. High altitude spectra were measured at sunrise and sunset so that long absorbing paths were able to be measured. Spectra were measured at 0.2 cm^{-1} resolution using a liquid nitrogen-cooled InSb detector, and by applying spectral deconvolution techniques to the measured spectra the resolution could be increased by about a factor of two. Mixing ratios of such gases as NO, NO_2, N_2O, CO, CO_2, CH_4, O_3, and H_2O were obtained; most of their results confirmed results determined previously by other methods, but a revised value for the mixing ratio of nitrous oxide in the stratosphere was able to be given. During this study, spectral absorption lines from NO were detected for the first time [47], in the stratosphere.

TABLE 10.2   Estimates of Pollutant Detectability Limits [38]

Pollutant	Best Measurement Frequency (cm^{-1})	C_{min} in Billionths of an Atmosphere (ppb)
Acetylene	735	0.02
Ammonia	965	0.2
Carbon monoxide	2180	1.0
Ethylene	950	0.2
Formaldehyde	2765	1.0
Formic acid	1100	0.8
Hydrogen chloride	2820	1.0
Hydrogen sulfide	1300	50.
Methane	3020	0.6
Nitric acid	880	0.6
Nitric oxide	1900	2.0
Nitrogen dioxide	1615	0.2
Ozone	1060	0.6
Peroxyacetyl nitrate	1160	0.2
Peroxybenzoyl nitrate	990	0.2
Phosgene	850	0.1
Propylene	915	1.0
Sulfur dioxide	1360	0.2

VII.   MATRIX ISOLATION SPECTROSCOPY

The application of matrix isolation techniques to far-infrared spectroscopy was described at the end of the previous chapter. Many applications of mid-infrared matrix isolation spectroscopy have been described in the literature for which grating spectrometers have been used, and in most cases there are few instrumental problems concerning energy limitation, measurement time or resolution. However, for samples dispersed in an inert gas matrix at a temperature close to 4 K the absorption lines can be very sharp. Indeed, the natural half-bandwidth of the bands is usually less than the resolution of many of the spectrometers which have been used for these measurements.

The resolution required to derive all the information contained in the spectra of most matrix isolated species is usually lower than about 1 cm^{-1}, but in the case of the analysis of molecules where different isotopes of one of the constituent atoms are present, high resolution measurements can yield important information.

One particularly interesting example of this type of study has recently been described by Gabelnick et al. [48–50], who measured the spectrum of a mixture of uranium oxides isolated in an argon matrix at fairly high resolution using a Fourier transform spectrometer. Urania, $UO_{2 \pm x}$, is an important component of nuclear fuel materials and its thermodynamic and vaporization properties have been extensively investigated. Gabelnick and his co-workers were able to derive important new data by condensing a mixture of argon and the uranium oxide vapors above heated urania onto a KBr plate at $15°K$, and measuring the high resolution spectrum of this matrix. Figure 10.26 shows a partial infrared spectrum of the matrix-isolated vapors above urania whose initial $U:O$ ratio was 2.5 and which was prepared using oxygen whose $^{16}O_2:^{15}O_2$ ratio was 3:2. Table 10.3 gives the observed vibrational frequencies and their assignments.

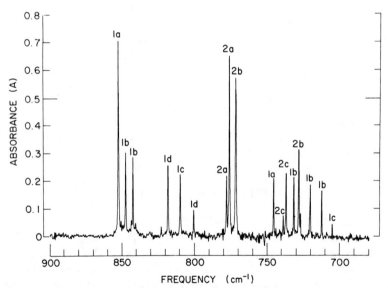

Fig. 10.26. Infrared spectrum of uranium oxide vapors above urania having an initial U/O ratio of 2.5 and prepared using a 3:2 mixture of $^{16}O_2:^{18}O_2$. The vapors were trapped in an argon matrix; the spectrum includes all observed peaks between 1500 and 450 cm^{-1}. (Reproduced from [48] by permission of the author and the North Holland Publishing Company; copyright © 1973.)

From their isotopic data, Gabelnick et al. were able to calculate that the O-U-O bond angle in UO_2 is approximately $180°$. From this computed near-linearity of the dioxide, the absence of v_1 for the pure isotopic UO_2 species (the symmetric stretching mode) was easily explained, and a value for this frequency could be calculated on the basis of the values of v_3 for $U^{16}O_2$, $U^{16}O^{18}O$ and $U^{18}O_2$.

It seems likely that similar studies can be made for other compounds whose thermodynamic properties are of interest and that the application of Fourier transform spectroscopy for the high resolution measurement of matrix-isolated species will become increasingly popular.

TABLE 10.3   Observed Uranium Oxide Vibrational Frequencies and Assignments

Frequency (cm^{-1})	Assignment	Designation in Fig. 10.27
776.10	$U^{16}O_2$ (v_3)	2a
771.70	$U^{16}O^{18}O$ ("v_3")	2b
737.05	$U^{18}O_2$ (v_3)	2c
728.35	$U^{16}O^{18}O$ ("v_1")	2b
852.50	$U^{16}O_3$ (v_3)	1a
847.55	$U^{16,18}O_3$	1b
842.50	$U^{16,18}O_3$	1b
818.35	$U^{16,18}O_3$	1b
809.90	$U^{18}O_3$ (v_3)	1c
800.65	$UO_3{}^a$	
745.60	$U^{16}O_3$ (v_1)	1a
731.80	$U^{16,18}O_3$	1b
720.50	$U^{16,18}O_3$	1b
712.60	$U^{16,18}O_3$	1b
705.15	$U^{18}O_3$ (v_1)	1c
820.00	UO	1d

^{a} Indicates uncertain assignment.

REFERENCES

1. W. J. Potts and A. L. Smith; *Appl. Opt.*, **6**, 257 (1967).
2. A. L. Smith and W. J. Potts, *Appl. Spectrosc.*, **26**, 262 (1972).
3. The Coblentz Society Board of Managers, *Anal. Chem.*, **38**(9), 27A (1966).
4. M. J. D. Low and S. K. Freeman, *J. Ag. Food Chem.*, **18**, 600 (1970).
5. P. R. Griffiths, unpublished work (1971).
6. R. O. Kagel and J. Baker, *Appl. Spectrosc.*, **28**, 65 (1974).
7. P. MacCarthy, H. B. Mark, and P. R. Griffiths, *J. Ag. Food Chem.* (in press, 1975).
8. D. L. Tabb and J. L. Koenig, Biopolymers (in press, 1975).
9. J. O. Alben, private communication (1972).

10. M. J. D. Low, A. J. Goodsel, and N. Takezawa, *Env. Sci. Technol.*, **5**, 1191 (1971).
11. A. J. Goodsel, M. J. D. Low, and N. Takezawa, *Env. Sci. Technol.*, **6**, 268 (1972).
12. M. J. D. Low, A. J. Goodsel, and H. Mark, in *Molecular Spectroscopy 1971*, P. Hepple, Ed., Inst. of Petroleum, London, 1972.
13. C. J. Percival and P. R. Griffiths, *Anal. Chem.*, 47, 154 (1975).
14. J. L. Koenig and D. L. Tabb, Can. Res. Develop., **7**, 25 (1974).
15. S. S. T. King, *J. Ag. Food Chem.*, **21**, 256 (1973).
16. P. R. Griffiths and F. Block, *Appl. Spectrosc.*, **27**, 432 (1973).
17. Harshaw Chemical Company, Crystal & Electronic Products Dept., 6801 Cochran Road, Solon, Ohio, 44139.
18. A. T. Stair, Aspen Int. Conf. on Fourier Spectrosc., 1970, G. A. Vanasse, A. T. Stair, and D. J. Baker, Eds., AFCRL-71-0019, p. 127.
19. M. J. D. Low, R. Epstein, and A. C. Bond, *Chem. Comms.*, p. 226 (1967).
20. M. J. D. Low, R. Epstein, and A. C. Bond, *J. Chem. Phys.*, **48**, 2386 (1968).
21. M. J. D. Low and S. K. Freeman, *Anal. Chem.*, **39**, 194 (1967).
22. M. J. D. Low, *Anal. Letters*, **1**, 819 (1968).
23. P. R. Griffiths and J. O. Lephardt, paper presented at the Pittsburgh Conf. on Anal. Chem. and Appl. Spectrosc., Cleveland (1969).
24. J. O. Lephardt and P. R. Griffiths, paper presented at the Pittsburgh Conf. on Anal. Chem. and Appl. Spectrosc., Cleveland (1970).
25. Wilks Scientific Corp., 140 Water St., Box 449, S. Norwalk, Ct., 06856.
26. M. J. D. Low, H. Mark, and A. J. Goodsel, *J. Paint Technol.*, **43**(562), 49 (1971).
27. L. S. Ettre, J. E. Purcell, and K. Billeb, *Perkin-Elmer Corp. Report* GC-AP-008 (1966).
28. Norcon Instruments Inc., 132 Water St., S. Norwalk, Ct., 06854.
29. D. Welti, *Infrared Vapour Spectra*, Heyden, London, 1970.
30. R. P. W. Scott, I. A. Fowlis, D. Welti, and T. Williams, Paper 20, Sixth Int. Symp. on Gas Chromatography, Rome (1966).
31. M. J. D. Low, *J. Ag. Food Chem.*, **19**, 1124 (1971).
32. K. L. Kizer, *Amer. Lab.*, p. 40 (June, 1973).
33. P. R. Griffiths, Chapter 7 in *Laboratory Methods in Infrared Spectroscopy*, 2nd Edition, R. G. J. Miller and B. C. Stace, Eds., Heydon, London, 1972.
34. Spectrotherm Corporation, 3040 Olcott Street, Santa Clara, Calif., 95051.
35. J. O. Lephardt and B. J. Bulkin, *Anal. Chem.*, **45**, 706 (1973).
36. H. Bar-Lev, *Infrared Phys.*, **7**, 93 (1967).
37. M. J. D. Low and H. Mark, *Water, Air and Soil Pollution*, **1**, 3 (1971).
38. P. L. Hanst, A. S. Lefohn, and B. W. Gay, *Appl. Spectrosc.*, **27**, 188 (1973).
39. J. U. White, *J. Opt. Soc. Am.*, **32**, 285 (1942).
40. E. R. Stephens, *Appl. Spectrosc.*, **12**, 80 (1958).
41. T. Hirschfeld, *FTS Notes*, Digilab Inc. (1972).
42. P. L. Hanst et al; E.P.A. *Preliminary Report* (January, 1974).
43. R. A. Schindler, *Appl. Opt.*, **9**, 301 (1970).
44. C. B. Farmer, 3rd FTS User Group Meeting, New York (November, 1973).
45. C. B. Farmer, *Can. J. Chem.*, **52**, 1544 (1974).
46. C. B. Farmer, O. F. Raper, R. A. Toth, and R. A. Schindler, 3rd Conf. on the Climatic Impact Assessment Program (February, 1974).

47. R. A. Toth, C. B. Farmer, R. A. Schindler, O. F. Raper, and P. W. Schaper, *Nature*, **244**, 7 (1973).
48. S. D. Gabelnick, G. T. Reedy, and M. G. Chasanov, *Chem. Phys. Letters*, **19**, 90 (1973).
49. S. D. Gabelnick, G. T. Reedy, and M. G. Chasanov, *J. Chem. Phys.*, **58**, 4468 (1973).
50. S. D. Gabelnick, G. T. Reedy, and M. G. Chasanov, *J. Chem. Phys.*, **59**, 6397 (1973). (1973).

MID-INFRARED REFLECTANCE SPECTROSCOPY

I. SPECULAR AND DIFFUSE REFLECTANCE

The demands of infrared reflectance spectroscopy are often quite different from those of absorption spectroscopy. Sometimes the sample being studied has a low reflectance or scatters the incident light so badly that only a very small fraction is able to reach the detector. In other studies, very accurate reflectance values are required for metal samples that reflect greater than 95% of the incident radiation across the complete spectrum. Another type of problem for which reflectance techniques have been used involves the detection of trace amounts of materials adsorbed on metallic surfaces. Each of these problems has been approached using Fourier transform spectroscopy, and each time useful new data have been collected.

The problem of measuring the reflectance of samples that *scatter* the incident radiation was one of the first to be attacked by Fourier transform techniques and is still producing interesting new results. For these measurements, the reflected light would usually be easily measured if the sample were perfectly flat, but an uneven surface causes most of the light to be reflected in other directions than toward the detector. The radiation that is reflected from a sample toward the detector in the same path that would be followed if it were perfectly flat is called the *specularly reflected* component, while the radiation that is reflected along different pathways is called the *diffusely reflected* component.

Most spectrometers will only measure the specular component of the radiation reflected from a given sample. This is the case if a sample is held in most commercial reflectance attachments for infrared spectrophotometers. If the sample scatters the incident radiation to any great extent the detector only picks up only a small fraction of the total reflected radiation so that spectra of rather low signal-to-noise ratio can result. Several types of samples that fall into this category have been measured by interferometry.

Much interest has been given to the feasibility of obtaining geological information from reflectance measurements. A feasibility study was carried out by Lyon [1] to investigate the reflectance spectra of a large number of rocks and minerals using a grating spectrometer. The samples studied had fairly large, flat, well-polished surfaces in order that the specular reflectance was maximized and the full reflected energy reached the detector. Lyon's evaluation of infrared spectroscopy for the compositional analysis of lunar and planetary soils established the feasibility of near-normal reflection for the purpose, but he suggested that the use of reflection techniques may have to be confined to manned laboratories since few naturally occuring rocks are flat or well polished. Low [2] measured the reflection spectra of small, irregular, unpolished samples using a low-resolution Block Engineering interferometer and showed that significant differences in the spectra of most rocks and minerals could be seen in the spectra. His experimental arrangement was quite simple (Fig. 11.1), and enabled good quality spectra to be obtained in

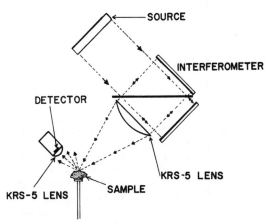

Fig. 11.1. Experimental arrangement used by Low to measure the reflectance spectra of rocks and minerals. (Reproduced from [2] by permission of the author and the Optical Society of America; copyright © 1967.)

short measurement times. For samples having relatively smooth surfaces, the spectral structure is quite dependent on the orientation of the sample (Fig. 11.2), although significant changes in band positions were not observed with microcrystalline, porous, or granular samples, such as D in Fig. 11.2 From the quality of these spectra it does not seem unreasonable to expect that infrared reflectance spectroscopy could be used, for example, in an unmanned spacecraft for the remote analysis of planetary surfaces.

Another study where the use of Fourier transform spectroscopy has enabled chemical information to be derived from a difficult sample involved the reflectance from plain paper, paper with a lacquer finish and colored paper with both pigments and lacquer present [3]. The reflectance from these samples was very low when measured using a microspecular reflectance attachment, but many significant bands could be seen in the spectrum

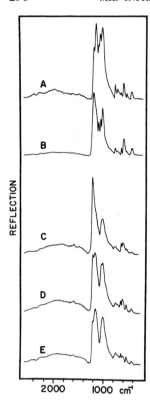

Fig. 11.2. Effect of orientation on reflectance spectra: *A.* reflection spectrum from one face of microline; *B.* reflection spectrum from the same face after 90° rotation; *C.* reflection spectrum from Na-plagioclase; *D.* reflection spectrum from the same face after 90° rotation; *E.* reflection spectrum from coarse-grained powdered Na-plagioclase. (Reproduced from [2] by permission of the author and the Optical Society of America; copyright © 1967.)

(Fig. 11.3). Similar types of sample were measured using the Willey Total Reflectance Spectrometer [4], and the maximum total reflectance was found to be greater than 50% showing that the diffusely reflected component from these samples is far greater than the specularly reflected component.

In practice it is found that the interpretation of diffuse reflectance spectra can be quite difficult at times, for two principal reasons. First, some bands that are symmetrical in the absorption spectrum can become somewhat distorted when seen in reflection. In addition, the radiation can be multiply reflected around the surface of diffusely reflecting samples before passing to the detector, which leads at times to rather unexpected results.

A comparison between spectra measured by diffuse reflectance and spectra measured using attenuated total reflectance (ATR) techniques [5] shows that each method has advantages and disadvantages. For example, the minimum surface area that can be studied using commercial microspecular reflectance attachments is smaller than can be studied using micro-ATR plates; the area of the sample whose spectrum is shown in Fig. 11.3 was less than 9 mm^2. Also

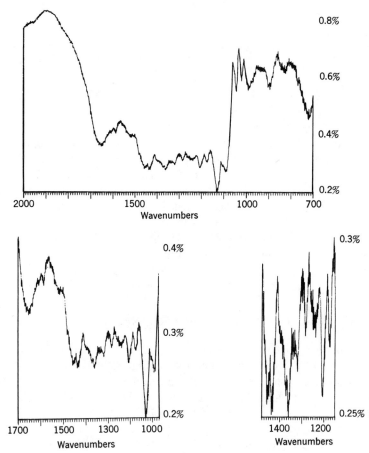

Fig. 11.3. Reflectance spectrum of a green ink on paper measured with a Digilab FTS–14 spectrometer using the Wilks Model 45 micro-specular reflectance attachment. The area of the sample under study was 9 mm²; each spectrum is plotted over decreasing frequency ranges and expanded to the same height. In the upper spectrum the ordinate range is from 0.8 to 0.2% and in the last spectrum it is from 0.30 to 0.25%. (Reproduced from [3] by permission of Heyden & Son Ltd.; copyright © 1972.)

some hard intractable samples, such as rocks and minerals cannot be studied by ATR since large areas of contact between the sample and the crystal cannot be attained. On the other hand, the specular component of the radiation from diffusely reflecting samples may be so small that spectra may only be measured with difficulty, even with a Fourier transform spectrometer, so that if good contact can be made between the sample and the ATR element, most spectroscopists will use ATR in preference to other types of reflectance techniques.

Another measurement presenting quite different problems from the previous types of sample is the accurate absolute determination of the reflectance of metals and other highly reflecting samples. These data can be extremely important, for example, heat transfer calculations in engineering. In studies with this type of sample, it has been found that a critical factor in determining the accuracy to which measurements may be made is the variation of the sensitivity of the infrared detector across its surface, and flux-averaging devices have to be used to give the most accurate results [6]. Using an integrating sphere in conjunction with an interferometer, Dunn and Coleman [7] were able to demonstrate very high accuracy reflectance measurements from metallic foil and mesh reflectors. When the Willey Total Reflectance Spectrometer was used for the measurement of samples of high refractive index and low absorption, such as KRS–5 [4], the sum of the reflectance and the transmittance of the samples was found to be equal to unity across the complete spectrum, which again indicates the usefulness of the integrating sphere in accurate measurements of the optical constants of solids.

II. ATTENUATED TOTAL REFLECTANCE SPECTROSCOPY

As a general rule it can be stated that interferometry is not well suited to the measurement of ATR spectra when rectangular ATR elements of the type designed for measurements on grating spectrometers are used. Unless beam distorting optics such as the toroid used by Low in his GC–IR studies [8] (see Fig. 10.17) are designed or very large ATR crystals are used, the circular beam from an interferometer will usually overfill the entrance plane of the ATR element and some energy will be lost. As a result, whereas an ATR attachment will often show a transmittance as high as 40% using the grating spectrometer for which it was designed, the same unit in a Fourier transform spectrometer usually attenuates the energy by about 90% unless very great care is taken in the alignment procedure. Under the relatively high energy conditions encountered in an ATR experiment, grating spectrometers are often competitive in performance with interferometers, especially when the time taken for computing and plotting the spectrum is included in the comparison. However, if additional energy is removed during the experiment for any reason, the advantages of Fourier transform spectroscopy start to reassert themselves.

Two examples where good ATR spectra have been measured by interferometry are found when a polarizer is present in the beam and when very small ATR elements are used for microsampling. In both cases the transmission of an ATR unit is often reduced below 10% even for a grating spectrometer.

Jakobsen [9] has been using a Fourier transform spectrometer in this way to study interfacial effects by infrared spectroscopy. Figure 11.4a shows the polarized ATR spectrum of a film of stearic acid. When the polarizer is rotated through 90° and the spectrum measured under these conditions is ratioed against the previous spectrum, several (but not all) bands are still observed (Fig. 11.4b). This spectrum suggests that the molecules of stearic acid are becoming ordered at the surface of the germanium ATR crystal. Jakobsen has measured the ATR spectra of a series of carboxylic acids and aliphatic alcohols down to very low surface coverages, and preliminary results again

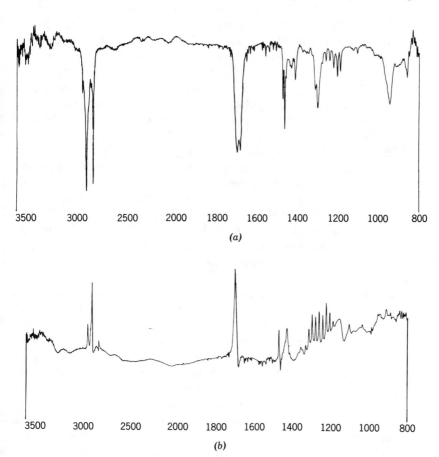

Fig. 11.4.   (a) Polarized ATR spectrum of stearic acid using a germanium crystal (45° polarization); the reference was the blank Ge crystal at the same polarization; (b) ATR spectrum of stearic acid on a Ge crystal measured using 45° polarization ratioed against the spectrum of the same sample measured using 135° polarization. (Measured on a Digilab FTS–14 spectrometer).

suggest that a considerable amount of ordering exists at the surface of the crystal and possibly even suggest that some chemisorption is occurring. The spectra of the molecules near the surface are reminiscent of the anionic forms of the sample being studied, even though the absorption bands due to the molecules only a few molecular layers distant from the surface are typical of the bulk sample measured by transmission techniques. The high signal-to-noise ratio of these spectra enabled Jakobsen to observe the weak bands at low coverages, whereas the same bands were lost in the noise when the same ATR plate was measured using a conventional grating spectrophotometer.

Low and Yang [10] have demonstrated that certain advantages can be derived from using a Fourier transform spectrometer for ATR spectroscopy which largely result from the flexibility of the data system of their spectrometer. They studied the ATR spectra of aqueous solutions of salts between 3500 and 2800 cm^{-1}, which is the region where the O-H stretching modes of water absorb. By storing a reference spectrum of water and using the ATR unit under identical conditions for the measurement of the solution spectrum, excellent compensation for the water band was achieved, as shown in Fig. 11.5. Low and Yang showed that the presence of almost any ionic solute produced a band at approximately 3200 cm^{-1}, a fact which they assigned to an alteration of the structure of liquid water. The contours of the water band appear to change depending on the nature and concentration of the solute. A band at 3200 cm^{-1} had previously been observed in spectra of aqueous solutions of ammonium salts which had been assigned to a vibrational mode of the $NH_4{}^+$ ion; in light of their results, Low and Yang have reassigned the spectra in this region, suggesting that the band at 3050 cm^{-1} is the asymmetric stretching mode of the ammonium ion while the band at 2875 cm^{-1} is due to a combination.

The same authors also applied ATR techniques to the measurement of the spectra of aqueous solutions below 2000 cm^{-1}. In a study of the ATR spectra of solutions of nitrate and nitrite ions using a Fourier transform spectrometer [11], they found that it was possible to analyze binary mixtures with $[NO_3{}^-] = 0.03\ M$ and $[NO_2{}^-] = 0.05\ M$. For aqueous solutions of organic compounds [12], the detection limit was found to be in the region of 0.01 M, while recognizable spectra could usually be measured with concentrations of approximately 0.05 M. These workers used an incidence angle of approximately 60° in order to obtain equivalent path-lengths of about 3 μm at 3450 cm^{-1} and 8 μm at 1500 cm^{-1}. A comparison between this work and studies of aqueous solutions made using transmission techniques with a Fourier transform spectrometer suggests that for the spectral region below 3000 cm^{-1} transmission techniques give the more sensitive results. Above this frequency, the capability of ATR to give very small penetration depths makes it relatively easy to find absorption bands beneath the strong O-H

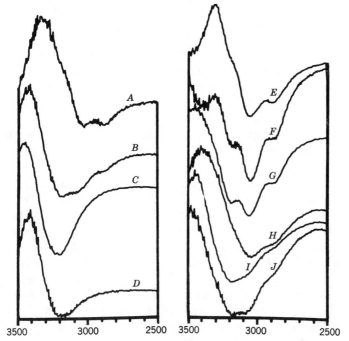

Fig. 11.5. Ordinate scale-expanded spectra of aqueous solutions measured by ATR using a Digilab FTS–14 spectrometer: *A.* 2 *M* NH$_4$NO$_3$ versus water; *B.* 2 *M* NH$_4$NO$_3$ versus saturated NaCl; *C.* Water versus saturated NaCl; *D.* 1.9 *M* NaClO$_4$ versus saturated NaCl; *E.* 5 *M* NH$_4$Cl versus water; *F.* 5 *M* NH$_4$Cl versus 1.9 *M* NaClO$_4$; *G.* 5 *M* NH$_4$Cl versus saturated NaCl; *H.* 1.6 *M* NH$_4$H$_2$PO$_4$ versus water; *I.* 1.6 *M* NH$_4$H$_2$PO$_4$ versus saturated NaCl; *J.* 1.6 *M* NH$_4$H$_2$PO$_4$ versus 1.9 *M* NaClO$_4$. (Reproduced from [10] by permission of the author and Marcel Dekker, Inc.; copyright © 1972.)

stretching bands in the water spectrum. Cells with a pathlength of only 3 μm are extremely difficult to construct for absorption spectroscopy, so that ATR is a very convenient technique for all measurements under very strong absorption bands.

III. REFLECTANCE TECHNIQUES FOR SURFACE STUDIES

The measurement of the spectra of coated or adsorbed samples on metal surfaces cannot strictly be referred to as reflection spectroscopy since the radiation that is being measured has actually passed *through* the sample twice and has been specularly reflected from the metal surface (Fig. 11.6). However the experimental techniques for studies of this type are much more closely

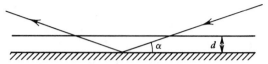

Fig. 11.6. To show that "reflectance" measurements of surface layers are in fact absorption measurements with a path length of $2d/\sin \alpha$.

allied to those of reflection spectroscopy than to those of absorption spectroscopy. Two types of samples have been studied interferometrically— relatively thick coatings such as those found on can surfaces and monolayer coverages of adsorbed gases. Measurement techniques for these two classes of material differ considerably in view of the different sample thicknesses.

Low and Mark [13] have described an experimental arrangement for measurements of can coatings which is very similar to the one used by Low [2] for measuring the reflectance spectra of rocks and minerals shown in Fig. 11.1. By reflecting the light from the interferometer off a small piece of the can and passing the light directly to the detector they were able to measure spectra from flat, crinkled, or generally odd-shaped pieces of metal cut from coated cans. For flat samples only a very few scans were needed to measure spectra of low noise-level and good repeatability, but for the deformed specimens more signal-averaging was needed.

Low and Mark then modified this arrangement so that spectra could be measured directly from the can. Light that had passed through the interferometer was focused and passed into the can and reflected into an Infratec detector which is a cylinder about 4 mm in diameter and 15 mm in length mounted in a glass tube (Fig. 11.7). This detector could be positioned deep inside the can which was held so that a portion of its inside surface was near the focus of the infrared beam. The can and detector were then moved around to maximize the signal. Finally, a multiple reflection technique was designed so that spectra of thinner layers could be measured. Figure 11.8 shows spectra of a thin film of spray enamel before and after heating. Spectrum C was measured after most of the film had been evaporated and only a blue translucent layer remained.

This work demonstrated the ease with which spectra of coatings can be measured but did not really show any increase in the sensitivity of measurements of weak absorption bands since the strongest bands in all the spectra shown by Low and Mark always absorbed at least 30% of the incident radiation. For the measurement of the spectra of *monolayers*, multiple reflection techniques almost invariably have to be used. Several theoretical treatments of infrared absorptions by thin layers on reflecting metals have been published. Greenler's treatment [14] most closely approximates the conditions

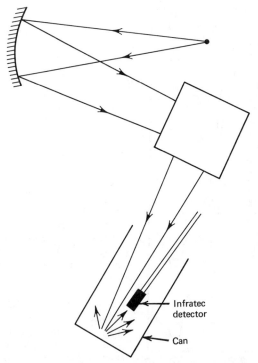

Fig. 11.7. Experimental arrangement used by Low and Mark to measure the spectra of can coatings nondestructively. (Reproduced from [13] by permission of the author and the Federation of Societies for Paint Technology; copyright © 1970.)

for a chemisorbed layer on a metal reflector; he considered the situation as a boundary value problem using the optical properties of the bulk metal and the properties of the liquid or solid state for the chemisorbed layer. These results were expressed as an infrared "absorption factor" for both the parallel and perpendicular components of polarization as a function of the angle of incidence. The treatment predicted that absorption by the parallel polarized radiation should be a maximum near grazing incidence and, at this optimum angle, the perpendicular absorption should be only 10^{-5} times as strong as the parallel.

In a short communication, Low and McManus [15] reported observing an absorption band at 2090 cm^{-1} due to carbon monoxide chemisorbed on platinum foil using a multiple reflection cell and a low resolution interferometer. This frequency is in good agreement with reports of the C-O stretching frequency for adsorbed carbon monoxide measured using transmission

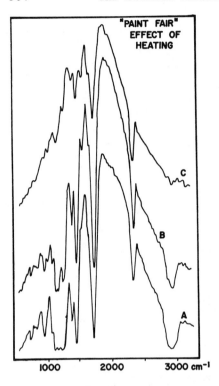

Fig. 11.8. Spectra showing the effect of heating surfaces treated with a clear spray enamel; spectra were measured with a Block Engineering Model 196 interferometer and are all single beam: *A*. spectrum of steel treated with enamel; *B*. spectrum of the same sample after heating to 200°C for several minutes; *C*. spectrum of the same sample after heating to 300°C for 10 min. (Reproduced from [13] by permission of the author and the Federation of Societies for Paint Technology; copyright © 1970.)

techniques. However, no estimate of the surface coverage was given in this paper. An examination of the reflection cell used for this work allows one to estimate about five reflections with an average incidence angle of about 70°. This geometry would suggest that the surface coverage was several molecular layers thick for a measurable absorption band to be measured.

In a much more detailed study, Harkness [16] was able to demonstrate absorption bands due to partial monolayer surface coverages as low as 5% for carbon monoxide on palladium. A multiple reflection cell consisting of many layers of palladium foil was designed so that multiple reflections can be achieved even though the beam entering the cell was highly collimated and the angle of incidence was close to 90° (grazing incidence). The collimated beam was passed through the interferometer and entered the reflection cell so that the angle of incidence of the beam on the foil was able to be well defined (Fig. 11.9). As a result of the amount of adsorbed impurity molecules found in commercially available palladium foil, very careful preliminary treatment of the surface was necessary. It was found that unless a pure metal surface was used, the amount of chemisorbed carbon monoxide that could be detected spectroscopically was negligibly small. The palladium was therefore oxidized,

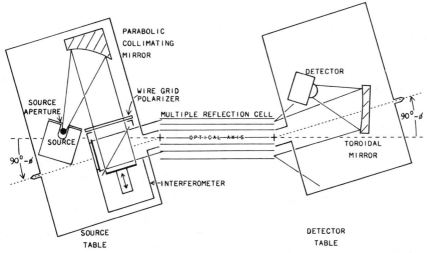

Fig. 11.9. Optical arrangement used by Harkness [16] to measure the spectra of partial monolayers of adsorbed gases.

reduced, and baked-out repeatedly before any carbon monoxide was allowed to come into contact with it.

Harkness found that the transmission of parallel polarized light through the cell was slightly different than that predicted by theory for close to grazing incidence due probably to either a slight inefficiency of the polarizer or a misorientation of the foil sheets. He also found that the absorption band due to adsorbed carbon monoxide was essentially independent of the state of polarization of the beam, in contrast to Greenler's prediction and other experimental studies. Harkness explained these deviations from theory in terms of a rough reflecting surface instead of the smooth surface required by the theory. A surface roughness of between 25 to 50 Å would be sufficient to randomize the dipole orientation with respect to the plane of incidence and cause this effect.

Harkness not only found a band (at about 2060 cm^{-1}) due to the C-O stretching vibration of chemisorbed carbon monoxide but also several other bands at lower frequency using the same beamsplitter. Although some of these bands were found in a later study [17] to be due to species other than chemisorbed carbon monoxide, one rather broad, weak band could be distinguished at 375 cm^{-1}, which was assigned by Apse to the Pd-C stretching vibration.

Although transmission studies using the finely divided metal in a refractory support have often [18–20] shown the band due to the C-O stretching vibration between 2050 and 2100 cm^{-1}, these studies represented the first time that any *low* frequency band had been observed, since the supports used for the transmission studies are opaque in the low frequency region, below 1300 cm^{-1}.

The studies of Harkness and Apse represent the only detailed interferometric investigation made on chemisorbed species, and even though they were made using a low resolution interferometer, they showed that Fourier transform spectroscopy is an extremely useful way of studying species adsorbed on metal surfaces. Since a collimated beam of radiation is desirable in this experiment (so that a uniform angle of incidence on the sample is achieved), an optical throughput which is less than the limiting value for high resolution measurements (Eq. 1.32) is required. Measurements can therefore often be made at a resolution sufficient to distinguish between the relatively broad bands due to the adsorbed molecules and the sharp vibration-rotation bands of the light gaseous molecules. The capability of distinguishing between gaseous and adsorbed molecules may become necessary when studying physically adsorbed molecules whose frequency shifts are much less than for chemisorbed species. In this respect Harkness noted that the only evidence for adsorbed CO when *impure* palladium was used as the substrate was an apparent reversal of the intensities of the P and R branches of the CO spectrum, presumably because the band due to the physically adsorbed carbon monoxide absorbed around the same frequency as the P branch of the gaseous molecule. Had he been able to use higher resolution for his experiments, Harkness could have distinguished between the bands due to gaseous and adsorbed carbon monoxide in the measurement.

In the general chemical analysis of surfaces by infrared spectroscopy, whenever bands due to any species at low coverage are to be measured, for example, by using a variable angle specular reflectance attachment, the angle of incidence should be made as high as possible. A considerable increase in absorbance is seen for bands due to surface species even on changing the angle of incidence from 45° to 75°. When a polarizer is placed in the beam so that only the parallel polarized component reaches the detector further increases in absorbance are observed. This author has found that invariably when surface species are to be detected by FTS, the measurements are best made at high incidence angle and using polarized radiation. This technique does have the disadvantage, however, of requiring that the sample has a large surface area.

REFERENCES

1. R. J. P. Lyon, "Evaluation of infrared spectrophotometry for compositional analysis of lunar and planetary solid," *NASA Technical Note D-1871* (1963).
2. M. J. D. Low, *Appl. Opt.*, **6**, 1503 (1967).
3. P. R. Griffiths, Chapter 7 of *Laboratory Methods in Infrared Spectroscopy*, 2nd Edition, R. G. J. Miller and B. C. Stace, Eds., Heyden, London, 1972.

4. R. R. Willey, personal communication (1973).

5. J. P. Deely, R. J. Gigi, and A. J. Liotti, *Tappi* (*J. Tech. Assoc. Paper Pulp Ind.*) **49**, 57A (1966).

6. S. T. Dunn, "Flux averaging devices for the infrared," *NBS Technical Note 279* (1965).

7. S. T. Dunn and I. Coleman, AIAA Paper 66–415, presented at AIAA Fourth Aerospace Sciences Meeting (1966).

8. M. J. D. Low, A. J. Goodsel, and H. Mark, in *Molecular Spectroscopy 1971*, P. Hepple, Ed., Inst. of Petroleum, London, 1972.

9. R. J. Jakobsen, unpublished work (1974).

10. M. J. D. Low and R. T. Yang, *Spectrosc. Letters*, **5**, 245 (1972).

11. R. T. Yang and M. J. D. Low, *Anal. Chem.*, **45**, 2014 (1973).

12. M. J. D. Low and R. T. Yang, *Spectrochim. Acta*, **29A**, 1761 (1973).

13. M. J. D. Low and H. Mark, *J. Paint Technol.*, **42**(No. 71), 265 (1970).

14. R. G. Greenler, *J. Chem. Phys.*, **44**, 310 (1966); *ibid*, **50**, 1963 (1969).

15. M. J. D. Low and J. C. McManus, *Chem. Comms.*, p. 1166 (1967).

16. J. B. L. Harkness, Ph. D. dissertation, M.I.T. (1970).

17. J. I. Apse, Ph.D. dissertation, M.I.T. (1973).

18. R. P. Eischens and W. A. Pliskin, in *Adv. in Catalysis and Related Subjects*, Vol. 10, Academic Press, New York, 1958.

19. L. H. Little, *Infrared Spectra of Adsorbed Species*, Academic Press, New York, 1966.

20. M. L. Hair, *Infrared Spectroscopy in Surface Chemistry*, Marcel Decker, New York, 1967.

INFRARED EMISSION SPECTROSCOPY

I. INTRODUCTION

Compounds can be classified by infrared spectroscopy through the variation of their transmittance, reflectance or emittance with frequency. At the present time, measurement of the transmittance spectrum of a sample is the usual means of identifying a material; reflectance spectroscopy is used infrequently while emission spectra are hardly ever measured. One of the principal reasons for the unpopularity of infrared emission spectroscopy is the experimental difficulties involved, especially when the sample is only able to be heated to very slightly above room temperature. In this case the energy at the detector is very low compared to the energy from the incandescent sources used in absorption or reflection spectroscopy, so that instrumental sensitivity limited the applications of the technique until the development of Fourier transform spectroscopy.

There are other difficulties besides sensitivity encountered in the measurement of infrared emission spectra. The radiation emitted by, or reflected from, the surroundings can be of comparable intensity to the emission from the sample when the temperature of the sample is low, and some means of discriminating against this spurious radiation often has to be applied (see Chapter 5, Section II). Temperature gradients across the sample can lead to spectra that are difficult to interpret.

For *quantitative* analysis by infrared emission spectroscopy, the emissivity spectrum, $\varepsilon(\bar{v})$, of the sample must be determined. Methods of measuring the emissivity of a sample in a laboratory environment were discussed in Chapter 5, Section II. For any sample, the sum of its transmittance, $T(\bar{v})$, reflectance, $\rho(\bar{v})$, and emissivity, $\varepsilon(\bar{v})$, at any frequency, v, is approximately equal to unity:

$$T(\bar{v}) + \rho(\bar{v}) + \varepsilon(\bar{v}) = 1. \tag{12.1}$$

Since the reflectance of most samples with discrete emission spectra is low, to a good approximation we have that

$$\varepsilon(\bar{v}) = 1 - T(\bar{v}) \tag{12.2}$$

The absorbance of a sample, $A(\bar{v})$, is related to the absorptivity, $a(\bar{v})$, path-length, b, and concentration, c, by the Beer-Lambert law:

$$A(\bar{v}) = -\log_{10} T(\bar{v}) = a(\bar{v}) \cdot b \cdot c \tag{12.3}$$

so that the emissivity of the sample is related to its concentration by

$$c = \frac{-\log_{10} \{1 - \varepsilon(\bar{v})\}}{a(\bar{v}) \cdot b} \tag{12.4}$$

or
$$\varepsilon(\bar{v}) = 1 - 10^{-a(\bar{v}) \cdot b \cdot c}. \tag{12.5}$$

Although this is a good approximation for gases, many condensed phase samples are held on a low emissivity support, such as metals of high reflectance [1]. In this case, absorption and reflection due to the support have to be allowed for in the calculation of emissivity. Steger and Rasmus [2] recently published quantitative expressions for calculating the emissivity of samples that are held on metal supports.

Although laboratory measurements of emissivity spectra show relatively few practical chemical applications, the quantitative analysis of heated *remote* samples by infrared emission spectroscopy may become one of the most important areas in which interferometric spectroscopy is applied. It is well known that the sensitivity of Fourier transform spectrometers is such that qualitative data are readily obtainable even from sources close to ambient temperature (*vide infra*). However, in order to obtain quantitative results, the emissivity of the sample must be calculated from the raw, uncorrected spectrum. Since it is rare to find a temperature sensor at the source in a remote measurement, some method of determining the temperature of the emitting sample from the spectrum has to be used.

Before any other manipulations can be carried out on the spectrum, it must be corrected for the varying response of the spectrometer as a function of frequency, due to such factors as beamsplitter efficiency, detector response, and amplifier characteristics. The correction factor, $F(\bar{v})$, may be measured in the laboratory by measuring the spectrum of a black-body at a known temperature and dividing this spectrum by the calculated values of the black-body emission at each frequency. After $F(\bar{v})$ has been measured in this way, all emission spectra, $I_S(\bar{v})$, should be divided by this factor to give a spectrum, $E_S(\bar{v})$, which is independent of the instrument on which the measurement has been made. If an emitting background contributes to the spectrum of the sample being determined, its spectrum, $I_B(\bar{v})$, should be measured if possible and subtracted from $I_S(\bar{v})$, that is,

$$E_S(\bar{v}) = \frac{I_S(\bar{v}) - I_B(\bar{v})}{F(\bar{v})}. \tag{12.6}$$

The temperature of the sample must then be determined from the corrected spectrum, $E_S(\bar{v})$. There are several methods in which temperature can be estimated from a spectrum and the technique used depends to a large extent on the type of molecule being studied. For large molecules with several strong emission bands in the spectrum, at frequencies $\bar{v}_1, \bar{v}_2, \ldots, \bar{v}_j, \ldots, \bar{v}_N$, the intensity of each band is dependent on the absorptivity of the band, $a(\bar{v}_j)$, and the spectral energy density of a reference black-body having the same temperature, T, as the sample, $E_R(T, \bar{v}_j)$. If $a(\bar{v}_j)$ is known for each band, a series of N simultaneous equations can be formulated in which the only unknowns are temperature and concentration.

The spectral energy density of a black-body at temperature T is given by

$$u(T, \bar{v}_j) = \frac{8\pi h \bar{v}^3}{\{e^{hc\bar{v}_j/kT} - 1\}}$$
(12.7)

and the emissivity is given by Eq. 12.5 as

$$\varepsilon(\bar{v}_j, c) = 1 - 10^{-a(\bar{v}_j) \cdot b \cdot c}.$$
(12.8)

Thus the emitted energy at $\bar{v}_j$ is given by

$$E_S(\bar{v}_j) = u(T, \bar{v}_j) \cdot \varepsilon(\bar{v}_j, c).$$
(12.9)

The N equations for $\bar{v}_1, \ldots, \bar{v}_j, \ldots, \bar{v}_N$ may be solved simultaneously to give the best fit for T and c.

For small molecules, there are usually too few bands to allow the above method to be used; in this case the temperature dependence of the rotational fine structure of a vibrational band can be used to the same end. Emission spectra are usually measured at fairly low resolution in order that an adequate signal-to-noise ratio is attained; however the band-widths of lines in the vibration-rotation spectrum of light molecules in air may be less than 0.1 cm^{-1}. Thus the Beer-Lambert law is no longer valid [3] and more complex methods than the one described earlier for polyatomic molecules have to be invoked. In particular, it appears important to deconvolve the ILS from the spectrum, which is by no means a trivial operation when the full-width at half-height of the molecules being studied may be ten times less than that of the ILS. This problem is currently being studied in several laboratories, and methods by which the temperature and concentration of light gases can be estimated from their remotely measured emission spectra should be reported in the near future.

Thus, infrared emission spectroscopy, while not being commonly used at the present time, shows promise of application in several areas of chemical analysis. One important reason for this statement is the fact that the measurement of emission spectra has become relatively routine since the development of mid-infrared Fourier transform spectrometers.

II.   EMISSION SPECTROSCOPY OF MATERIALS IN CONDENSED PHASES

Many of the early measurements of the infrared emission spectra of liquids and solids were performed by Low and Coleman [4–7]. Their first paper [4] demonstrated the feasibility of characterizing organic compounds through their emission spectra measured using an early Block Engineering rapid-scanning interferometer. The samples were heated slightly above ambient temperature (30°C) and a room temperature thermistor bolometer was used as the detector. One drop of a liquid sample was placed on an aluminum surface and covered with a KBr flat. For most of the compounds measured in this way there was excellent correspondence between band frequencies observed in the infrared emission spectrum measured interferometrically and the absorption spectrum of the same material measured on a grating spectrometer (Fig. 12.1).

Low and Coleman [5, 6] have also demonstrated the application of infrared emission spectroscopy for the analysis of surface species. In one study [5] the emission spectrum of aluminum foil was measured, after which the sheet was wiped with a paper treated with oleic acid and then wiped "dry". The emission spectrum of this sample at 30°C showed three weak bands in the region of 1700 to 1600 cm^{-1}; the authors attributed these bands to the carbonyl stretching frequency of the undissociated acid and to the asymmetric and symmetric O-C-C vibtations of the oleate onion. Similar bands were observed when the spectrum was measured with the sample at 100°C, indicating that the oleic acid reacts with the aluminum surface to form an oleate.

Low [6] also published the results of a similar study with tricresyl phosphate on a steel surface. The emission spectra of untreated steel and steel treated with a layer of tricresyl phosphate were measured with the sample at 40°C. The expected bands due to tricresyl phosphate were observed superimposed on the relatively featureless spectrum of steel. The sample was then heated to 400°C for 5 min and then cooled down again to 40°C at which temperature its emission spectrum was once again measured. It was noted that the bands due to the P-O-C linkage had disappeared while some of the aromatic bands were still evident in the spectrum, suggesting that cleavage of the P-O-C linkage occurs in the tricesyl phosphate that remained on the surface and that some products containing P-O groups and aromatic systems remained bound to the steel surface even after heating to 400°C. These species were not identified by Low; however this paper demonstrates the potential usefulness of infrared emission spectroscopy for monitoring surface species. Apparently, nobody has ever compared the relative detection limits of reflectance and emittance measurements for surface studies.

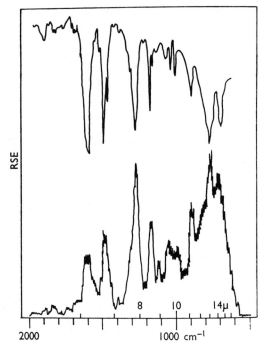

Fig. 12.1. One of the first emission spectra of chemical interest, that of aniline at 30°C, was measured with a low-resolution interferometer using 200 scans; the upper spectrum shows a transmittance spectrum of aniline for comparison. (Reproduced from [4] by permission of the author and the Chemical Society; copyright © 1965.)

In all the spectra shown in these three papers, the emission from the metal surface comprised a fairly large proportion of the total signal even though the emissivity of metal surfaces is small. The emitted radiation from the background was not subtracted from the spectrum measured after a layer of the sample had been applied since this work was performed in the very early days of chemical infrared Fourier transform spectroscopy when data manipulation was much more difficult than it is using today's versatile computerized instrumentation. It seems likely that remeasurement of the type of systems studied by Low using the higher resolution interferometers available today, in conjunction with background subtraction techniques, could yield important new data on surface reactions.

Another paper published by Coleman and Low [7] at about the same time as the three papers just described discussed the feasibility of microsampling by infrared emission spectroscopy, with particular reference to the identification of pesticides in the microgram range. Samples were prepared by depositing a known amount of a solution of known concentration onto a plate of NaCl or KBr using a microsyringe. The solution was allowed to air-dry after deposition, and discs of NaCl or KBr were then placed over the samples which were spread in thin layers in irregular patches varying from about 5 to 15 mm in diameter as a result of the evaporation of the solutions. The salt plates were then placed on a first-surface mirror which was positioned on a small horizontal hot-plate placed within a few millimeters of the entrance aperture of the vertical spectrometer. Apparently no focusing optics were used, so that the interferometer not only viewed the sample but also a fairly large area of the mirror where no sample was present.

Spectra of DDT measured in this way suggested that identifiable spectra could be measured with 100 μg of sample, but that only one or two very weak bands could be observed if only 10 μg was present. However, the spectra reproduced in this paper are of very mixed quality, and it might be possible to detect the same bands in a spectrum measured after only 1 μg of DDT had been applied. When Malathion and Dieldrin were studied at low levels, several bands could be seen in the infrared emission spectrum, and it seems likely that amounts of pesticides as small as 10 μg could be characterized in this way. This study was performed to demonstrate the feasibility of the technique and the spectra should not be taken to represent the ultimate now possible with more advanced equipment. The method of sample preparation does not seem at all suited for microsampling since the solution is allowed to spread over too large an area before evaporation. In addition it seems likely that if the detector viewed an image of the sample alone, rather that of a much larger area, considerably better detection limits may have been attained.

It has been suggested [8] that for certain applications infrared emission spectroscopy may be a more sensitive technique than absorption spectroscopy. This claim needs to be investigated further for applications of chemical interest. No comparision of spectra of the same sample measured using absorption and emission techniques have been made for either surface studies or microsampling, and a study of this kind appears to be long overdue.

Emission spectra can only be measured when there is a difference in temperature between the source and the detector. Generally the source is held at a higher temperature than the detector but there is no reason why the source should not be cooler than the detector. Low [9, 10] has studied the emission spectra of minerals using both techniques, first [9] with the sample (source) at 40°C and the detector at about 30°C and second [10] with the sample at 20°C and the detector at about 24°C. The experimental arrangement for the latter measurement is worth noting and is shown in Fig. 12.2. The box is made out of copper plate and is insulated with a 5-cm layer of

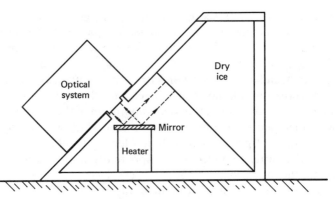

Fig. 12.2. The experimental arrangement used by Low and Coleman for measuring the emission spectra of samples below ambient temperature. (Reproduced from [10] by permission of the author and the Optical Society of America; copyright ©️ 1966.)

Styrofoam. The compartment on the right was filled with dry ice and the interior of the left compartment was sprayed with a nonreflecting black paint and could be flushed with nitrogen. The samples were crushed to obtain better temperature control. If a first surface mirror is placed on the heater, the radiation emitted by the detector system is reflected to, and absorbed by, the walls of the box. Because the walls are nonreflecting and the mirror is a weak emitter, the radiation returning to the detector is minimal. The spectrum measured in this way very closely approximates the true emission spectrum of the detector.

The procedure used by Low and Coleman to correct for this emission was to first measure the interferogram with the sample on the heater. The sample is then replaced by a mirror and successively larger numbers of interferograms are subtracted. Figure 12.3 shows the effect of this procedure. The relative spectral emission (RSE) of the subtracted spectrum shows more and more discrete spectral structure until after approximately 250 scans the emission spectrum of the sample, granite, is measured with no background contribution from the emission of the detector.

Very high quality emission spectra of mineral samples at low temperature were measured in this way. Measuring infrared emission spectra with the sample cooler than the detector has several advantages. First, the sample does not have to be heated so that thermal decomposition does not occur. The most important advantage is that the effects of temperature gradients in the sample are reduced. When the source is at a higher temperature than the detector, it is usually heated from below the emitting surface. Thus the lower surface has a higher temperature than the upper surface. Radiation emitted

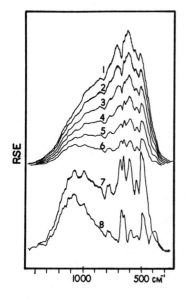

RSE

1000   500 cm⁻¹

Fig. 12.3. The emission spectrum of granite shown to illustrate the method of background subtraction necessary for measuring the spectra of samples below ambient temperature; 200 scans from the sample were measured. The spectrum number and the total number of scans of detector emission subtracted were: *1*, 0; *2*, 50; *3*, 100; *4*, 150; *5*, 200; *6*, 250; *7*, 250; *8*, 300. Spectra 7 and 8 are shown under greater ordinate scale expansion than the other spectra (Reproduced from [10] by permission of the author and the Optical Society of America; copyright © 1966.)

from below the upper surface can therefore be absorbed before it reaches this surface which can lead to anomalous features in the emission spectra of bulk samples.

Griffiths [1] has discussed this effect with reference to two spectra of calcite (Fig. 12.4); one was the emission spectrum of a fairly thin film on an aluminum sheet (after Kagel [11]) and the other that of a bulk sample (after Low [9]). Both were measured with the sample at a higher temperature than the detector. The most intense band in the infrared spectrum of calcite is found at 1400 cm⁻¹. The spectrum of the *film* of calcite shows most of this band as it would be expected in an emission spectrum; however in the central region of the band there is a decrease in the measured emission giving the band the appearance of a doublet. In this region the absorption of the radiation emitted below the surface of the film by cooler molecules near the surface is greater than the emission of the molecules near this surface. In the *bulk sample* this effect is so great that the band appears to be an absorption on a general black-body background; in this spectrum there appears to be a small "emission" maximum at the center of the band.

If samples are being measured with their temperature greater than that of the source it is obviously preferable to use a thin film of the sample rather than a crushed bulk solid or a thick film. This effect was illustrated by Griffiths [1] who measured the emission spectra of thick and thin films of silicone grease on an aluminum sheet (Fig. 12.5). In the spectrum of the thin film, the characteristic bands of the silicone grease are easily seen whereas the

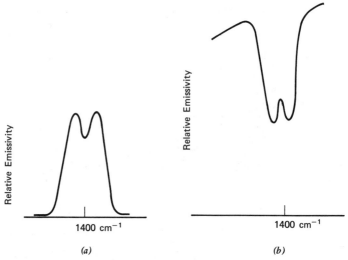

(a) (b)

Fig. 12.4. Emission spectra of (A) a fairly thin layer of calcite deposited on aluminum foil, after Kagel [11], and (B) a bulk sample of calcite, after Low [9]. (Reproduced from [1] by permission of the Society for Applied Spectroscopy; copyright © 1972.)

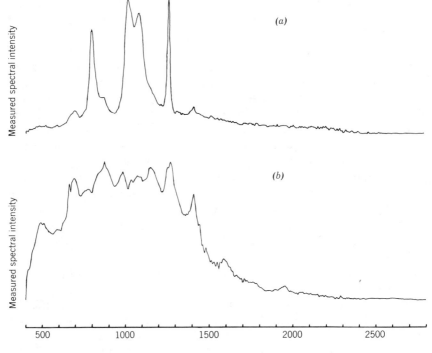

Fig. 12.5. The emission spectra of (A) a thin layer and (B) a thicker layer of silicone grease on heated aluminum foil. (Reproduced from [1] by permission of the Society for Applied Spectroscopy; copyright © 1972.)

general emission in the spectrum of the thick film masks the discrete band structure.

If the sample is held at ambient temperature this type of effect cannot occur so easily and a much better approximation to the true emission spectrum should be measured. Low and Coleman measured the emission spectrum of a bulk sample of the mineral steatite both with the sample above and below the detector temperature (Fig. 12.6a and b). The spectrum measured with the sample at 250°C shows a black-body background with the bands due to the mineral seen in "absorption" in the same way that the spectrum of bulk calcite appears in Fig. 12.4b. When the temperature of the sample was held below the detector temperature, all the bands are seen in emission. Thus the latter technique for the measurement of emission spectra yields a much better approximation to the emissivity of a bulk sample than if the sample is hot and the detector is cool.

The spectra shown in Fig. 12.6b are not corrected for the variation in relative spectral emission with frequency for a black-body at this temperature. Thus, strong bands at, say, 1000 cm^{-1} appear to be less intense than bands of lower emissivity at lower frequency, say 500 cm^{-1}, because of the fact that the energy emitted from a black-body at 20°C is greater at 500 cm^{-1} than at 1000 cm^{-1}. One method of getting a better estimate of the emissivity of a sample is to replace the sample by a sheet of black-painted aluminum at the same temperature. Since the emission spectrum of this sheet closely approximates that of a black-body, by ratioing the emission spectrum of the sample against the emission spectrum of the blackened sheet a good approximation to the emissivity spectrum can be obtained.

One other early paper on infrared emission spectroscopy is worthy of note. In 1966, Low [12] published the emission spectrum of *in vivo* human skin. The emission spectra of normal skin and skin after any fat on the surface was washed off with acetone showed slight differences. More spectacular differences were observed when a layer of oleic acid was applied to the surface of the skin and then washed off with acetone; further differences were noted after the skin was abraded with sandpaper. However no explanation of the differences was attempted and the spectra seem to be in conflict with the type of spectrum which one would expect for this kind of surface. Low has suggested that infrared emission spectroscopy could be used for the nondestructive examination of skin in studies of pigmentation, the nature of infected areas, the presence or absence of secretions and similar effects. No work subsequent to this early paper has been reported.

III. EMISSION SPECTROSCOPY OF GASES

One of the areas where infrared emission spectroscopy shows its greatest potential is in the remote identification and quantitative determination of

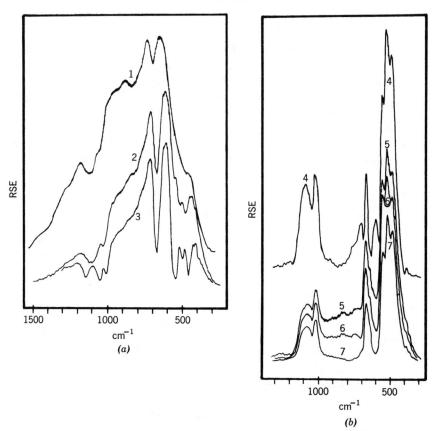

Fig. 12.6. Emission spectra of the mineral steatite measured (*A*) with the sample at 250°C and (*B*) with the sample at 20°C and the detector at 24°C, measured using the experimental arrangement shown in Fig. 12.2. Spectrum *A.1* is the spectrum of the steatite sample at 250°C; the spectra in (*B*) were measured by subtracting varying numbers of background scans from the spectrum measured with the steatite sample at 20°C and the detector at 24°C. It is apparent how much more spectral information can be obtained with the sample below ambient temperature. (Reproduced from [10] by permission of the author and the Optical Society of America; copyright ⓒ 1966.)

heated gases, and in particular the remote sensing of pollutants in stack gases. The feasibility of these determinations has been demonstrated by Low and Clancy [13], who used an 8-in. reflecting telescope to collect radiation emitted from the plume of a smokestack located approximately 200 yards from their interferometer.

The spectrum of the sky background near the smokestack was measured first (Fig. 12.7*a*); radiation transmitted through the atmospheric window between 1100 and 700 cm^{-1} is prominent in this spectrum. The telescope was then pointed at the plume and a second spectrum was measured (Fig. 12.7b) which revealed several prominent bands not observed in the background spectrum. On subtraction of the sky background spectrum from the spectrum of the plume plus the sky, the spectrum of the plume alone was obtained (Fig. 12.7c). Among the bands in this spectrum, strong features due to SO_2 are seen, together with bands due to CO_2. Note the absorption at 670 cm^{-1} of the strong CO_2 emission line which is caused by absorption of the radiation emitted at this frequency by cooler molecules in the atmospheric path to the detector. The origin of another emission band (marked f in the spectrum) is uncertain, but may well be caused by incompletely burned fuel.

These data are qualitative in nature and a considerable amount of further work was indicated for quantitative results to be obtained from this type of spectrum. Very recently two papers were published where a more rigorous approach to the study of infrared emission spectroscopy for the determination of the concentration of components of stack gases was outlined.

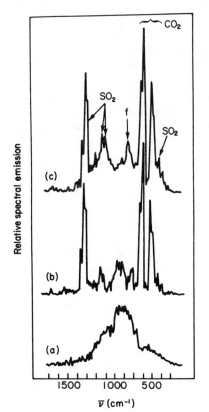

Fig. 12.7. Infrared emission spectra of a smoke-stack effluent, each spectrum measured using 100 scans. *A.* sky near plume; *B.* plume plus sky; *C.* plume only (spectrum *A* subtracted from spectrum *B*). (Reproduced from [13] by permission of the author and the American Chemical Society; copyright © 1967.)

Prengle et al. [14] used an interferometer for the quantitative determination of the components of stack gases and obtained results which were in remarkably good agreement with data taken simultaneously by other methods (in-stack monitoring and mass spectroscopy of a collected sample). One drawback of their method involves the way in which the temperature of the effluent gas was determined; the amplitude of the interferogram at zero retardation was used as a measure of the temperature, apparently without correcting for the radiation emitted by the sky background. By comparing the amplitude of the interferogram with the reading of a thermocouple mounted inside the stack, Prengle and his co-workers were able to derive a calibration curve from which the temperature of the stack gas was able to be estimated to an accuracy of $\pm 10°C$. It was claimed that their calibration curve could be used for all stacks of that particular "class" but that recalibration would be needed if a stack of a different class was to be monitored. The presence of particulate matter in the effluent will strongly affect the total radiation being monitored at the detector, so that it is not likely that the temperature could be measured accurately by Prengle's method unless the effluent was free of particulate matter (which the stack on which he obtained his data was).

Chan et al. [15] have described what should be a more general method for estimating the temperature. They noted that the rotational fine structure of a vibrational band is strongly temperature dependent, and proposed a method whereby the temperature of a gas can be estimated by examining the contour of emission bands. The temperature of the gas should be able to be correlated with the frequency at which the emission is a maximum for any given band. They make the point that it is not necessary to completely resolve the fine structure of any emission band in order to determine the frequency at which the maximum intensity occurs, but rather that only the contour need be measured. However, since the band-width of the fine structure varies from molecule to molecule (being wider for lighter molecules) and since the resolution needed to just resolve this fine structure is also strongly dependent on the molecule being monitored, it is difficult to envisage how this method can be used for spectra measured interferometrically as a general method for monitoring several different pollutants; rather it can be used in the determination of one particular species using an individual sensor. It appears to this author that in order to use Fourier transform spectroscopy for the remote sensing of effluent gases, the measurements should be carried out at the highest resolution consistent with an adequate signal-to-noise ratio, so that the fine structure of each band can be adequately analyzed for temperature measurement, quantitation and elimination of interferences.

Prengle et al. [14] have not given any experimental details concerning either the detector which they have used for their measurements or the resolution at which their spectra were measured in their remote sensing measurements, neither do they show any spectra; thus a realistic appraisal of their results is difficult. Their quoted results are in excellent agreement with data from other sensors present in the stack which they were monitoring. CO, NO, NO_2, and saturated and unsaturated hydrocarbons were measured at levels of between 10 and 10,000 ppm to an accuracy of $\pm 28\%$. These results undoubtedly show the feasibility of using infrared Fourier transform spectroscopy for the quantitative remote sensing of stack gases. However, the method of Prengle's group is not sufficiently general to be universally applied to monitoring all types of smokestacks. In particular, no account was taken of reabsorption of radiation emitted in the stack by cooler molecules in the path to the detector. For certain compounds that are present in the atmosphere at relatively high concentration (e.g., CO_2 and CO), this effect provides a source of error that may be well outside the uncertainty due to the sensitivity of the equipment or fluctuations in the composition or temperature of the stack.

Several groups are currently working on the problems found in remote sensing, and it is likely that a more general method than that of Prengle will be proposed in the reasonably near future.

IV.   FAR-INFRARED EMISSION SPECTROSCOPY

The intensity of a given emission line is dependent on the Boltzman population of the excited state for the transition of interest. The principal reason for the weakness of most mid-infrared emission bands is the large difference in the energy of the excited and the ground states such that at ambient temperature the concentration of species in the excited state is rather low. If rotational transitions occurring in the far-infrared region of the spectrum could be used for monitoring gaseous species through their emission spectrum, the increased concentration of molecules in excited states should facilitate the task of measuring their emission spectrum.

However, far-infrared emission spectroscopy does not appear to provide a viable alternative to mid-infrared measurements for the remote sensing of pollutants at ground level in view of the severe attenuation of the emitted radiation by atmospheric water vapor. On the other hand, the concentration of water vapor in the upper atmosphere is very low, and far-infrared interferometry has been used to measure the concentrations of several species in the stratosphere.

The first measurements of the very far-infrared ("submillimeter") emission spectra of species in the stratosphere were made by Gebbie and his coworkers [16], who showed that the spectrum consists of a large number of pure rotation lines, principally due to H_2O, O_2, and O_3. Quantitative measurements of water and ozone could be made both because of the high signal-to-noise ratio of the spectrum and because the presence of lines due to O_2 could be used to "internally calibrate" the effective air mass of the spectrum, since the mixing ratio of O_2 to air is known to be constant in the stratosphere.

Harries and his co-workers [17–18] have subsequently incorporated a helium-cooled InSb bolometer in place of the Golay detector used by Gebbie and found an improvement of more than an order of magnitude in sensitivity over the previous measurements. With this improvement many new and much weaker emission lines were able to be observed, due not only to ozone but to other species such as NO_x, HNO_3, and possibly SO_2. These measurements are of particular importance since at this time the concentration of these species in the stratosphere has become of great interest in view of the possible effects to the environment caused by emissions from the supersonic airliners currently being developed. It seems ironic that these latest measurements were taken from just such an aircraft.

An NPL "cube" interferometer was installed in a Comet 2E for the first measurements [19] and for subsequent measurements the equipment was installed in a prototype Concorde airliner [18]. Most measurements were taken at the maximum resolution of the interferometer $(0.06\ \mathrm{cm}^{-1})$ and phase modulation techniques were used to further increase the sensitivity of later measurements.

A typical spectrum is shown in Fig. 12.8. The depression of the zero atmospheric emission level is due to the property of the Michelson interferometer that part of the radiation emitted from the detector enters the interferometer and is reflected back to the detector exactly out-of-phase with the spectrum of the source [17]. Being broad and featureless, the reflected detector spectrum depresses the baseline but does not cause any difficulty in the measurement of the sharp emission lines. All the features in this spectrum could be assigned to (at least) one previously observed emission line from one of the species thought to be present, and several lines could be caused by several different transitions or their blend. The lines marked with asterisks are probably due to HNO_3, NO_2, N_2O, or (possibly) SO_2.

Concentrations were determined in two ways. For water and ozone lines, ratioing the O_2 line equivalent widths gave a mixing ratio directly [20]; however, this method is not applicable to the weak lines of the nitrogen and sulfur compounds. For these species laboratory intensities obtained both experimentally and theoretically have been used to derive their atmospheric concentration [18]. The atmosphere is considered as an equivalent path to the laboratory case and a corresponding amount of emitting gas is derived. A correction is applied for the lower temperature of the gas in the atmosphere and a resultant total amount (in units of cm-atm) is obtained which can be converted into the volume mixing ratio.

A difficulty was found in obtaining really accurate data because so many

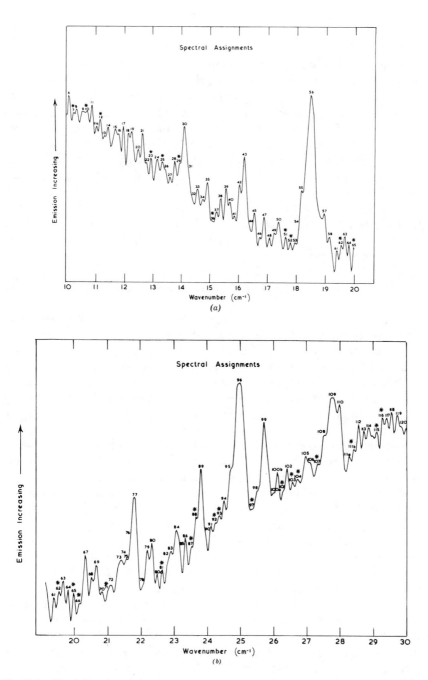

Fig. 12.8. Far-infrared emission spectrum of the stratosphere measured with a phase-modulated NPL cube interferometer with an InSb detector, mounted in a Concorde airliner. The lines marked by asterisks are those used for determining the concentrations of HNO_3, NO_2, N_2O, and SO_2; resolution is 0.067 cm^{-1}. (Reproduced from [18] by permission of the author; Crown copyright is reserved.)

parameters that could affect the measurement varied simultaneously, such as altitude, latitude and longitude, tropopause level and the time of day. In spite of these variations, the results of Harries [18] are in excellent agreement with other published results. Of the major species present in the stratosphere, ozone is present with a volume mixing ratio of 5 ± 1 ppm, while the corresponding value for water is 3 ± 0.3 ppm. The three nitrogen compounds that could be fairly accurately determined are HNO_3, N_2O, and NO_2 which were calculated to be present with a volume mixing ratio of 2.8 ± 0.7, 270 ± 50, and 22 ± 8 ppb, respectively. For N_2O the value above the tropopause decreased to 210 ± 30 ppb.

These results suggest that the use of far-infrared Fourier transform spectroscopy could be of great importance for upper atmosphere analysis and may give results for more species at greater accuracy than any other method currently being used.

V. INFRARED CHEMILUMINESCENCE

In recent years, infrared chemiluminescence has become an important tool for the study of reaction dynamics [21]. In measurements of this type, the infrared radiation emitted from the products of a chemical reaction is measured under conditions such that the vibrational relaxation of the products is effectively eliminated. Information obtained from infrared chemiluminescence experiments has been used to determine the distribution of reaction energy into the vibration, and sometimes the rotation and translation, of the products.

The usual method of arresting vibrational relaxation in these experiments is simply to work under conditions of very low pressure (10^{-4} torr or lower). At this pressure it has been calculated that two intersecting molecular beams will have a total flow rate of 100 μmole/sec, while the rate of production of excited molecules may be as low as 1 μmole/sec. The intensity of radiation emitted under these conditions which can be measured on a spectrometer has been estimated in the following way [22]. The collection optics of a typical spectrometer will usually have a collection efficiency of the order of 1%, which therefore means that the molecule flux which is studied in a chemiluminescence experiment is about 10 picomoles/sec, or 5×10^{15} molecules/sec. Since the average residence time in the spectrometer beam is of the order of 3×10^{-4} sec and the average radiative lifetime is approximately one second, it has been calculated that the photon flux at the detector for a chemiluminescence experiment is about 10^{12} photons/sec. If a spectrometer with a throughput of 10^{-2} cm^2-sr is used for the measurement, the photon flux is further reduced to 10^7 photons/sec.

At these very low levels of incident radiation, extremely sensitive infrared detectors have to be used, and Moehlmann et al. [22] have analyzed the situation occurring when a liquid helium cooled mercury-doped germanium bolometer is used to detect infrared chemiluminescence of this intensity. They have shown that if the ambient materials are all at room temperature, the principal source of noise is the statistical fluctuations of the background radiation hitting the detector; at 300 K this background is 4×10^{18} photons/cm/sec. By reducing the temperature of the surroundings to 77 K, the background radiation is reduced by a factor of 10^6, and at this level the noise equivalent photon flux (c.f. Chapter 8, Section I) is approximately 5×10^5 photons/sec. Under these conditions the signal-to-noise ratio for a chemiluminescence experiment of the type described above would be of the order of 20, assuming that the *complete* spectrum is viewed for one second.

In previous experiments of this type, only very low resolution data have been measured using dispersion spectrometers, since a prism or grating spectrometer large enough to give a throughput of 0.01 cm^2-sr and giving a resolution of, say, 1 cm^{-1} would be far too large to cool to 77 K. Moehlmann et al. therefore modified a Michelson interferometer so that it could be cooled with liquid nitrogen, as described in Chapter 4, Section IV.E; using this system they have been able to measure infrared chemiluminescence spectra at reasonably high resolution from the reaction of fluorine atoms with a variety of monosubstituted ethylene compounds.

A beam of fluorine atoms, generated by the thermal dissociation of F_2 at high temperature, was reacted with a beam of $CH_2 = CHX$ molecules, where $X = H$, CH_3, Cl, or Br. Two types of reaction mechanism are possible for these molecules, abstraction reactions such as:

$$F + C_2H_3X \rightarrow HF + C_2H_2X$$

and substitution reactions of the kind:

$$F + C_2H_3X \rightarrow X + C_2H_3F.$$

Thus, substitution reactions for each of the reactants produce vinyl fluoride, together with a H, Cl or Br atom or a CH_3 radical.

Published spectra for these reactions are shown in Fig. 12.9. It can be seen that each spectrum has three common bands, which have been assigned to vinyl fluoride by comparison with its absorption spectrum. The spectrum for which propene was used as a reactant shows an additional band at 1280 cm^{-1} which has been assigned to emission from the methyl radical. The spectrum from the ethylene reaction displays a broad weak band centered at 1250 cm^{-1} which cannot be assigned to vinyl fluoride; this band appears to be due to the C_2H_3 radical formed from the abstraction reaction occurring at the same time as the substitution reaction.

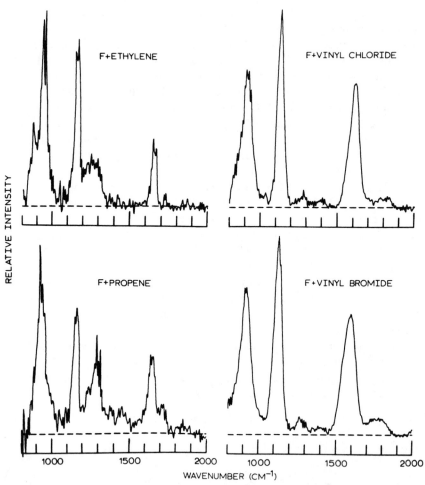

Fig. 12.9. Normalized emission spectra from the four reactions studied by Moehlmann et al. The zero intensity for each spectrum is indicated by the dotted line. (Reproduced from [22] by permission of the author and the American Institute of Physics; copyright © 1974.)

Analysis of these spectra has allowed the way in which the reaction energy is partitioned over the various vibrational modes to be determined. The results indicate that for the reactions of vinyl chloride and vinyl bromide, the reaction energy was statistically partitioned over the vibrational modes of the vinyl fluoride product, while the partitioning for the reactions in which ethylene and propene were the reactants was non-statistical. The difference in energy partitioning for these reactions was attributed to the existence of a potential energy barrier in the exit channel of the reaction coordinate for the

H and CH_3 elimination reactions. This conclusion verifies earlier theoretical predictions that for H and CH_3 elimination reactions such a barrier should be of the order of 3 and 8 k-cals/mole, respectively, while for the elimination of halide atoms, the barrier was predicted to be virtually nonexistent.

This experiment was the first infrared chemiluminescence measurement to be made in which at least one of the products was not a diatomic molecule. Previous measurements of infrared chemiluminescence generally had used filters to analyze the emitted radiation, so that only very low resolution information could be obtained. The experiments of Moehlmann et al. were the first measurements of this nature in which relatively high resolution spectra could be gathered across the entire infrared region, thereby allowing the calculations of the partitioning of reaction energy to be made for polyatomic molecules. It is quite apparent that the use of a cooled Michelson interferometer for this experiment was the principal reason that spectra of such a high signal-to-noise ratio were able to be measured. It seems very probable that other reactions will be studied using Fourier transform techniques now that the feasibility of the method has been demonstrated.

VI.   "TIME-RESOLVED" EMISSION SPECTROSCOPY

In all the applications that have been covered so far, it has been assumed that the intensity of the source is constant or the absorbance of each absorption band does not vary significantly throughout the measurement. On the other hand, it is possible to envisage an application of Fourier transform spectroscopy in which the spectrum of a source of rapidly varying intensity can be measured as a function of time or the nature of an intermediate formed in a rapid reaction can be determined.

Suppose that the lifetime of a certain species is less than one second; measurement of its spectrum by conventional dispersive techniques would be difficult but could be achieved in the following manner. If a particular species may be rapidly generated using a pulsed source, its emission across a certain frequency element could be examined using a monochromator and measuring the emitted radiation at several time intervals after the pulse. The next frequency element may be measured after letting the system relax down to the ground state and the pulsing once again to reexcite the molecules. The emitted energy across this new frequency element would again be measured as a function of time. Performing this operation for each spectral frequency element would allow the complete spectrum to be measured.

The same type of measurement could be envisaged for an interferometric measurement of the same set of spectra so that the multiplex advantage of interferometry can be realized. Consider an interferometer with a stepping motor drive used in place of the monochromator of the previous experiment.

The source is pulsed with the movable mirror of the interferometer stationary at a known retardation, and the emitted intensity is measured as a function of time. The movable mirror is then stepped forward to the next position, the source is repulsed and the signal is then measured once more as a function of time. This procedure would be continued until the end of the interferogram. On transforming the interferogram using a set of data points taken at identical times after each pulse, the spectrum at this time is obtained. The transform of each set of such data points allows a set of spectra of the source at different times after the pulse to be calculated.

Murphy and Sakai [23] have described an experiment in which this procedure was carried out for the purpose of investigating the relaxation of vibrationally excited CO_2. The near resonance between the energy levels of N_2 at 2331 cm^{-1} and CO_2 at 2340 cm^{-1} produces a very efficient transfer of vibrational energy from the excited $N_2^\ddagger$ molecule to the CO_2 molecule. Carbon dioxide may be vibrationally excited by mixing it with nitrogen which has been activated by passing it through a microwave cavity. It will decay either by radiation or by deactivation at the walls or quenching by another molecule. The schematic of set-up for the experiment is shown in Fig. 12.10.

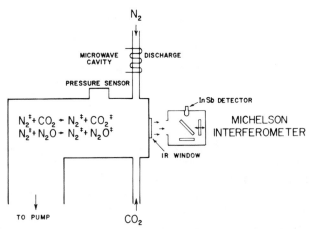

Fig. 12.10. The experimental arrangement used by Murphy and Sakai. (Reproduced from [23] by permission of the author.)

The microwave discharge was pulsed so that active N_2 was produced, and the radiation emitted from the activated CO_2 as a result of the energy transfer:

$$N_2^\ddagger + CO_2 \rightarrow N_2 + CO_2^\ddagger$$

was measured with a liquid-nitrogen cooled InSb detector using a d.c. amplifier. It was found that for any given position of the interferometer mirror, the signal varied with the time after the pulse in the manner shown in Fig. 12.11a, while the spectra computed from the interferograms measured after the time indicated by 1, 2, 3, and 4 are shown in Fig. 12.11b. From these data Murphy and Sakai were able to show that the typical vibrational temperatures achieved were of the order of several thousand degrees Kelvin, while the rotational temperature remains at about 350 K since rotational relaxation occurs rather quickly (typically 3–10 collisions).

While "time-resolved" interferometry has been used here for an experiment of physical chemical importance, there appears to be no reason why a similar technique should not be applied for the chemical analysis of species which are formed as intermediates in rapid chemical reactions, possibly using emission techniques but more probably by absorption spectroscopy.

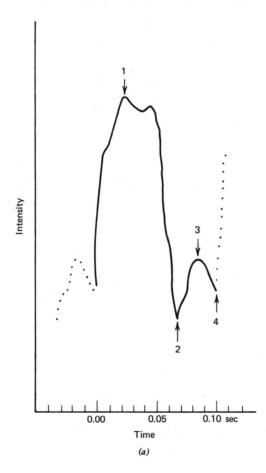

(a)

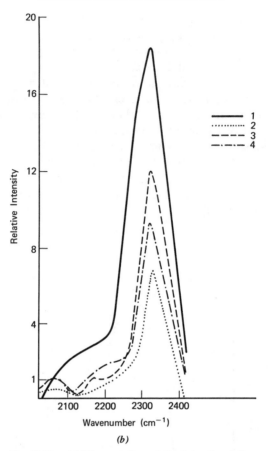

Fig. 12.11. (a) The variation of the intensity of the signal at the detector for a single excitation; interferogram samples were taken at the four positions marked; (b) the transform of the interferograms taken from the four sampling positions shown in (a). (Reproduced from [23] by permission of the author.)

REFERENCES

1. P.R. Griffiths, *Appl. Spectrosc.*, **26**, 73 (1972).
2. E. Steger and D. Rasmus, *Appl. Spectrosc.*, **28**, 376 (1974).
3. D. A. Ramsay, *J. Am. Chem. Soc.*, **74**, 72 (1952).
4. M. J. D. Low, L. Abrams, and I. Coleman, *Chem. Comms.*, p. 389, (1965).
5. M. J. D. Low and I. Coleman, *Spectrochim. Acta*, **22**, 369 (1966).
6. M. J. D. Low, *J. Catalysis*, **4**, 719 (1965).

7. I. Coleman and M. J. D. Low, *Spectrochim. Acta*, **22**, 1293 (1966).
8. G. Wijntjes, 2nd Fourier Transform User Group Meeting, Cleveland (March, 1973).
9. M. J. D. Low, *Nature*, **208**, 1089 (1965).
10. M. J. D. Low and I. Coleman, *Appl. Optics*, **5**, 1453 (1966).
11. R. O. Kagel, unpublished work (1970).
12. M. J. D. Low, *Separatum Experientia*, **22**, 262 (1966).
13. M. J. D. Low and F. K. Clancy, *Env. Sci. Technol.*, **1**, 73 (1967).
14. H. W. Prengle et al., *Env. Sci. Technol.*, **7**, 417 (1973).
15. S. H. Chan, C. C. Lin, and M. J. D. Low, *Env. Sci. Technol.*, **7**, 424 (1973).
16. H. A. Gebbie, W. J. Burroughs, J. E. Harries, and R. H. Cameron, *Astrophys. J.*, **154**, 405 (1968).
17. J. E. Harries, N. R. W. Swann, J. E. Beckman, and P. A. R. Ade, *Nature*, **236**, 159 (1972).
18. J. E. Harries, NPL Report DES 16 (1972).
19. J. E. Harries and W. J. Burroughs, NPL Report DES 7 (1970).
20. W. J. Burroughs and J. E. Harries, *Nature*, **227**, 824 (1970).
21. T. Carrington and J. C. Polanyi, *MTP International Review of Science, Physical Chemistry*, **9**, 135 (1972).
22. J. G. Moehlmann, J. T. Gleaves, J. W. Hudgens, and J. D. McDonald, *J. Chem. Phys.*, **60**, 4790 (1974).
23. R. E. Murphy and H. Sakai, Aspen Int. Conf. on Fourier Spectrosc., 1970, G. A. Vanasse, A. T. Stair, and D. J. Baker, Eds., AFCRL-71-0019, p. 301.

THE FUTURE OF FOURIER TRANSFORM
SPECTROSCOPY

It is certainly true that the principal advantage of using a Fourier transform spectrometer rather than a grating monchromator for infrared spectroscopy is derived from the multiplex advantage of interferometry. Most of the published applications of FTS have described experimental measurements that would have been extremely difficult to make using a grating spectrophotometer. However, even a few years ago it was still widely believed that interferometry would not meet with a general acceptance within the chemical community, since so many chemists believed that the *practical disadvantages* of FTS outweighed its *theoretical advantages*.

The principal disadvantage of interferometry concerned the fact that the raw data are in the form of an interferogram which requires digitization before it may be transformed into a spectrum. A second, less fundamental, objection to FTS concerned its application to mid-infrared spectroscopy: several users of slow-scanning far-infrared interferometers attempted to modify their instruments for mid-infrared spectroscopy and obtained results that did not compare favorably with the corresponding measurements made on a grating spectrometer. Technological advances that occurred in the late 1960s and early 1970s have led to the development of techniques whereby *spectra* of all wavelengths can now be measured routinely, and many of the earlier objections as to the practicability of FTS no longer apply.

Let us briefly discuss how the various difficulties encountered with early interferometers have been overcome, and how the methods by which these problems have been solved have often presented the spectroscopist with rather unexpected secondary benefits.

First and foremost, it has been the commercial development of minicomputers and their associated mass memories, such as discs and tapes, which has enabled the Fourier transform of an interferogram to be performed on-line, without the need to interface in some way to a remote computer. Also of prime importance in this respect was the application of the Cooley-Tukey fast Fourier transform algorithm to infrared spectroscopy, so that computing time between the completion of the measurement of the interferogram and the start of the spectral plot has been reduced to a matter of seconds for low resolution spectra. Once the spectrum has been calculated, it is stored in the data system of the spectrometer and it can subsequently be operated upon in many ways, some of which were described in Chapter 6, Section IV. Indeed, several measurements which have been taken using a Fourier transform spectrometer were possible, not because of any great sensitivity advantage of the interferometer, but rather because of the ease with which the digital spectral data can be manipulated.

In several sections of this book we discussed why the development of *rapid-scanning* interferometers enabled the signal-to-noise ratio of mid-infrared interferograms to be reduced to the point that noise as well as signal could be sampled on digitization of the interferogram; signal-averaging then enables spectra to be obtained at any desired sensitivity. Until the development of small inexpensive helium-neon lasers, no rapid-scanning interferometer capable of measuring high resolution spectra had been constructed. The principal reason for this fact was that interferograms generated by the early instruments had to be sampled at equal intervals of time rather than equal intervals of retardation. Thus on signal-averaging, successive interferograms lost phase coherence after scanning less than one millimeter. When the output from a laser is used to generate a sinusoidal reference signal, the main interferogram can be sampled at equal intervals of retardation over very long optical path differences. In this way fairly high resolution spectra are able to be measured using a rapid-scanning interferometer.

The use of lasers as devices for digitizing the interferogram at precisely equal intervals of retardation also leads to an additional, less obvious benefit to the spectroscopist. Since the frequency of the laser is constant, each output point in the spectrum is at a constant precise fraction of the frequency of the laser. Thus the frequency of each data point in the spectrum is always the same, even in spectra measured several days apart, and there is no danger that numerical operations performed between these spectra will be invalid because of unexpected instrumental frequency shifts. This is particularly important when two single-beam spectra are ratioed to obtain a transmittance spectrum. If the reference spectrum shows sharp spectral features (due, for example, to atmospheric water vapor) and a small frequency shift occurs between the time the reference and sample spectra are measured, it is essentially impossible to completely compensate for the sharp background features.

We noted in Chapter 1 that there is a small frequency shift that occurs when the beam passing through the interferometer is not perfectly collimated. This shift may be allowed for by adjusting the value of the laser frequency entered in the computer program for the FFT until the measured frequencies of calibrant bands correspond exactly to the values given in standard reference data. In the same way, spectra can either be plotted with respect to vacuum or air wavenumbers merely by changing the value of the laser reference frequency in these programs.

Finally, an important factor that at one time was regarded as a major disadvantage of FTS was the cost of the instrumentation. Even at the beginning of 1973 it was not possible to buy a computerized mid-infrared Fourier transform spectrometer for under $60,000. However the rapid advances in the technology of solid-state devices has enabled the cost of the data system to be reduced so that even before the end of 1974 the price of some spectrometers with disc data systems had been reduced below $40,000. Thus the difference in the cost of grating and Fourier transform spectrometers is getting smaller each year.

In light of the technological developments in the recent past, is it possible to predict future commercial developments in FTS? We have seen that the first instruments to be manufactured were the mechanically simple slow-scanning far-infrared interferometers made by RIIC and Grubb-Parsons in England and the rapid-scanning unreferenced interferometers made by Block Engineering in America. Later the more sophisticated computerized far-infrared interferometers were developed by RIIC (after their acquisition by Beckman Instruments), by Coderg in France and by Polytec in Germany, while the first automated laser fringe-referenced mid-infrared instruments were developed by Digilab in America. For a couple of years the only Fourier transform spectrometers that were commercially available were general purpose instruments, but recently several special purpose instruments have also been developed. For example, the Willey Model 318 Total Reflectance Spectrometer is designed primarily for measuring the reflectance spectra of diffusely reflecting surfaces. The Spectrotherm ST−10 spectrometer is specifically designed for the on-line measurement of the infrared spectra of gas chromatographic peaks. The EOCOM Model 7200 system has been designed for the quantitative analysis of trace gases in the atmosphere using a long-path cell.

Can we expect any further developments along these lines? Although there do not appear to be any major new instruments on the immediate horizon, it seems to this author that there may be one or two areas that are potentially amenable to the design of a special purpose Fourier transform spectrometer. One of these might be an instrument for the on-line identification of components of mixtures separated by a high-pressure liquid chromatograph; another might concern the monitoring of smoke-stack effluents.

In view of the rapidity with which the instrumental scene is changing in so many different areas, it may also be asked whether there are any techniques which may be developed in competition to FTS. Another multiplex method for infrared spectroscopy is Hadamard transform spectroscopy (HTS). There are good theoretical reasons [1] why HTS should be less convenient for chemical spectroscopy than FTS, and no results of chemical significance have yet been published for which a Hadamard transform spectrometer has been used. For *absorption* spectroscopy, the use of tunable lasers has been shown to give very sensitive results when high resolution spectra are required. As yet, however, the range over which infrared lasers can be tuned is not sufficient to allow the complete mid-infrared spectrum to be measured on one system. In addition, the *rate* at which they can be tuned is too slow to permit tunable lasers to be used for many of the applications for which FTS has been successfully applied.

In conclusion, it can truly be said that FTS has provided the chemical spectroscopist with a powerful tool for measuring infrared spectra at higher sensitivity or at greater speed than is possible by the use of competitive techniques. It is equally certain that important new measurements will be made in many different areas of infrared spectroscopy using Fourier transform spectrometers.

REFERENCES

1. T. Hirschfeld and G. Wijntjes, *Appl. Optics*, **12**, 2876 (1973); *ibid*, **13**, 1740 (1974).

INDEX